I0051270

African Journal of Science, Technology, Innovation and Development (AJSTID)

Vol. 1, No. 1, 2009

Contents

African Journal of Science, Technology, Innovation & Development

AJSTID

African Journal of Science, Technology, Innovation and Development (AJSTID)

Vol. 1, No. 1, 2009

Copyright 2009©AJSTID
Adonis & Abbey Publishers Ltd
ISSN: 2042-1338 (print)
ISSN: 2042-1346 (online)

Cover Design: Asia Iftikhar
Layout Artist: Jan B. Mwesigwa

Adonis & Abbey
Publishers Ltd

Adonis & Abbey Publishers Ltd
P.O. Box 43418
London, UK
SE11 4XZ

http://www.adonis-abbey.com
Also visit the journal's website:
http://www.ajstid.com

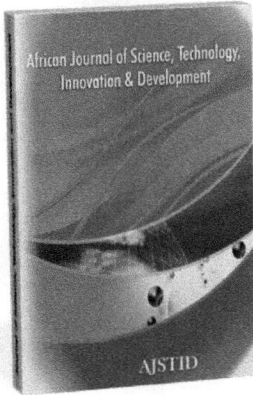

AJSTID is a multi-disciplinary and refereed international journal on science, technology, innovation and development in Africa and other low-income countries. It was established in appreciation of the role and importance of innovation in the development processes and the recognition of the relative absence of research in this area in Africa. AJSTID encourages research along the following broad areas:

- Role of science, technology and innovation in the processes of industrial growth and development.
- Emergence and the making of innovation systems in the context of broader socio-economic development.
- Exploring the inclusion of innovation and knowledge in subnational, regional, global and local innovation networks and cross border integration processes in Africa.
- Research on the interaction among governments, industries, businesses, universities and communities in the use and application of science, technology and innovation policies; particularly comparative works that have implications for all developing economies.
- Critiques of science, technology and innovation policies and their applications in Africa.
- Exploring co-evolutions, broad-based innovations and indigenous knowledge systems in the context of African development.

Submitting Papers

Authors are required to submit original papers, that is, papers submitted should not have been previously published or be currently under consideration for publication elsewhere. All papers are refereed through a double blind process. You may send an electronic copy of your paper (MS Word file attached to an e-mail) to:

Dr. Angathevar Baskaran, Email: anga_bas@yahoo.co.uk and
Prof. Mammo Muchie, Email: mammo@ihis.aau.dk

For detailed guidelines: visit **http://www.adonisandabbey.com/**
Also visit: http://www.ajstid.com

African Journal of Science, Technology, Innovation and Development
Vol. 1, No. 1, 2009
pp. 6-8

Introducing the New Journal: Notes from the Editors

Mammo Muchie and Angathevar Baskaran (Editors-in-Chief)

Early this year the Adonis-Abbey Publishers (London) invited us to start and be principal editors of a new African Journal on Science, Technology, Innovation and Development (AJSTID), and we agreed to take on that responsibility. After a lot of hard work and goodwill from different people – sponsoring institutions, reviewers, contributors, copy editors, and leading thinkers and scholars in this area, the first issue has come out in August 2009. We plan to produce the next two issues (combined) before the end of 2009.

The response from scholars and researchers from all over Africa and across the world to the call for contribution has exceeded our expectations. It demonstrated that there has been a real gap waiting to be filled. Researchers from every region in Africa and beyond have submitted scientific papers representing the work of both emergent researchers and senior researchers for peer review.

One of the core aims of AJSTID is to encourage emerging research scholars particularly in Africa to publish their work in a journal that will be read by the global community of scholars working on these topics. We also intend to encourage well established researchers from all over the world to contribute to the journal and help to build it to be one of the top ranking journals. Although the main focus of AJSTID is Africa, we encourage researchers and scholars from the rest of the world to contribute to the journal, particularly comparative works that have implications for all developing economies.

We decided that AJSTID should be distinct from many other journals in this area. For this, we included research notes/commentaries -- individual accounts from thinkers and experts in this area, and also accounts and contributions from research networks about their activities. The research notes/commentaries will provide on how accomplished researchers selected certain research directions and not others and what influenced them to make their choices to do certain research and not others. Accounts/ contributions from research networks and their activities are included to stimulate learning on how functioning research networks work. Another novel addition is to invite publication of reprints of path breaking research articles from highly rated journals such as *Research Policy*.

In this first issue of AJSTID, Christopher Freeman has contributed a succinct account of why he made a life-long commitment to work on the economics of innovation. Bengt-Åke Lundvall draws lessons from Danish development for innovation work in Africa. The CAAST-Net have written on their network that aims to enhance research cooperation between Europe and Africa. The IDRC funded a pan-African research on e- local goveranance that included 10 African countries. The research note by Timothy Mwololo Waema, Winnie Mitullah, and Edith Adera have generated a research framework for monitoring and evaluating local e-governance from the experience drawn mainly from Kenya.

Nelson and Winter's *'Useful theory of innovation'* that came out in 1977 in *Research Policy* is reprinted in this issue. We hope reprinting such seminal and classical work will stimulate emerging researchers and reinforce theoretical development in this area not only in Africa but also in wider global context.

The rest are all scientific papers and they were all contributions mainly by African emergent and senior researchers but also including researchers from outside Africa (Canada): Bitrina D. Diyamett from Tanzania, Michael Gastrow and Saahier Parker from South Africa, Abdelrasaq Al-Suyuti Na-Allah from Nigeria, Mammo Muchie from Ethiopia, Watu Wamae, Timothy Mwololo Waema, Winnie Mitullah, and Edith Adera from Kenya, Abdelkader Djeflat from Algeria, and Laxmi Prasad Pant and Helen Hambly Odame from Canada.

Finally the book reviews included four works that are selected for their power to amplify what is most needed to strengthen research, knowledge and innovation in Africa.

We wish to recognize here the support and encouragement of the Institute for Economic Research on Innovation (IERI) at Tshwane University of Technology (TUT), in South Africa to launch this journal. Without IERI's, TUT's and the South African Research Chairs Initiative's (SARChI's) support, this journal would not have seen the light of the day.

There is hard and very tedious work involved in producing any new journal and more so in this case. In Africa such a journal is needed more than perhaps anywhere in the world. The statistics on scientific publication, as indeed on many other indicators when it comes to Africa, remain always negative and disconcerting.

AJSTID wishes to break this by providing opportunity to publish so that in the long run science, technology, innovation and development become a power to strengthen research capacity, research systems and the African research and knowledge area.

AJSTID hopes to be a positive force in helping to attract the best researchers and the best research to counter the negative projection that Africa has suffered for so long.

It is with pleasure we invite the readers, institutions, and contributors to use this journal and build it as one of the means to make Africa acquire agency and freedom in running its knowledge and research life for generations to come.

We are proud and indeed privileged to be the facilitators to the birth of this research journal that will contribute to building Africa to emerge also as a research, knowledge and innovation area.

Mammo Muchie & Angathevar Baskaran
Editors -in- Chief

African Journal of Science, Technology, Innovation and Development
Vol. 1, No. 1, 2009
pp. 9-52

In Search of Useful Theory of Innovation

Richard R. Nelson* and Sidney G. Winter**

Abstract

This essay presents an overview of the prevailing theoretical literature on innovation, probes the adequacy of existing theory to guide policy regarding innovation, and sketches some directions for more fruitful theorizing. The focus is on the vast interindustry differences in rates of productivity growth, and other manifestations of differential rates of technological progress across industries. It is argued that the most important policy issues involve finding ways to make the currently lagging sectors more progressive, if in fact that can be done. Theory, to be useful, therefore must organize knowledge and guide research regarding what lies behind the uneven performance of the different economic sectors. In fact prevailing theory cannot do this, for two basic reasons. One is that theory is fragmented, and knowledge and research fall into a number of disjoint intellectual traditions. The second is that the strongest of the research traditions that bear on the differential innovation puzzle, research by economists organized around trying to 'fit' production functions and explain how production functions 'shift', neglects two central aspects of the problem; that innovation involves uncertainty in an essential way, and that the institutional structure supporting innovation varies greatly from sector to sector. The bulk of the paper is concerned with sketching a theoretical structure that appears to bridge a number of presently separate subfields of study of innovation, and which treats un-certainty and institutional diversity centrally.

Key Words: innovation theory, productivity, profit maximization, R&D strategy, economic development
JEL Classification: O31, O32, O34, O4

*   Institution of Social and Policy Studies, Yale University, New Haven, Connecticut 06520, USA

**   Economics Department, Yale University, New Haven, Connecticut 06520, USA

1. Introduction

This essay presents an overview of selected aspects of prevailing theoretical understanding of innovation, and attempts to sketch some directions that would seem fruitful to follow if we are to achieve a theoretical structure that can be helpful in guiding thinking about policy. We are using the term innovation as a portmanteau to cover the wide range of variegated processes by which man's technologies evolve over time. By a theory we mean a reasonably coherent intellectual framework which integrates existing knowledge, and enables predictions to go beyond the particulars of what actually has been observed. It seems apparent that if scholarly knowledge is to be helpful to deliberation about possible policy directions, theory must be wide enough to encompass and link the relevant variables and their effects, and strong enough to give guidance as to what would happen if some of these variables changed.

It is also apparent that, in trying to design policies aimed at so complex a web of social phenomena as innovation, intelligent policy makers are going to look to the scholarly community for advice. Indeed, research by economists and other social scientists on various aspects of innovation has had a major impact on recent policy thinking. In the United States, the Congressional hearings on science and technology policy, and the speeches of high officials on these matters, are full of 'facts' discovered by economists about the major role of technological advance in productivity growth, as a source of comparative advantage, etc. The Joint Economic Committee of the United States Congress recently commissioned a review of the literature. The economists' concept of externality is much bandied about in the policy dialogue in the UK as well as the US, and is a stock part of the rhetoric emanating from OECD.[1]

The current dialogue regarding policy toward innovation rests on two premises. The first is that technological advance has been a powerful instrument of human progress in the past. The second is that we have the knowledge to guide that instrument toward high priority objectives in the future. The first premise is unquestionable: the latter may be presumptuous. While all the attention recently given by politicians to scholars is flattering, we believe that the scholarly community has much less to say about appropriate policy toward innovation than many scholars like to believe. Prevailing theory of innovation has neither the breadth nor the strength to provide much guidance regarding the variables that are plausible to change, or to predict with much confidence the effect of significant changes.

Implicit in these critical remarks is a set of commitments on our part as to the nature of the major policy issues involving innovation. In a nutshell they are these.

First, at the present time the appropriate policy objectives are not well characterized as more effective general stimulation of technological advance, but involve more selective targeting. In the past technological advance has been extremely uneven across economics sectors and industries. Table 1, drawn from Kendrick's recent study, displays the vast intersectoral variations in growth of total factor productivity, and labor productivity, that were experienced over the 1948-1966 period (Kendrick, 1973). Evidence of great intersectoral differences of productivity growth was provided earlier in Kendrick's 1961 study (Kendrick, 1961), and still earlier in work by Salter (1966) and Schmookler (1952). The phenomenon seems to obtain in other countries as well as the US. Where other measures or indices of technological progress are available, they correlate reasonably well with the productivity measures, and show a similar cross industry dispersion.

Table 1: Productivity growth in the private business economy by industry group and industry, 1948-1966.

	Percentage yearly change in total factor productivity	Percentage yearly change in output per worker
Private domestic business economy	2.5	3.1
Farming	3.3	5.6
Mining	4.2	4.6
Metal	2.4	2.9
Coal	5.2	5.8
Crude petroleum and natural gas	3.2	2.3
Nonmetallic mining and quarrying	2.6	3.2
Contract construction	1.5	2.0
Manufacturing	2.5	2.9
Nondurables	2.6	3.2
Food, except beverages	3.0	3.4
Beverages	2.2	2.9
Tobacco	1.1	2.7
Textiles	4.0	4.3
Apparel	1.9	2.2
Paper and paper products	2.5	3.0
Printing	2.7	2.7
Chemicals	4.9	6.0
Petroleum refining	3.0	5.5
Rubber products	3.9	4.0
Leather products	1.7	1.7
Durables	2.4	2.8
Lumber	3.5	3.9
Furniture	2.9	2.9
Stone, clay, glass	2.4	3.2
Primary metal products	1.6	2.1
Fabricated metals	1.9	2.2
Machinery, except electric	2.6	2.7
Electric machinery	3.7	-4.1
Transportation equipment and ordinance	3.2	3.2

Table 1: continued from previous page

	Percentage yearly change in total factor productivity	Percentage yearly change in output per worker
Instruments	2.9	3.7
Miscellaneous manufacturing	3.5	4.0
Transportation	3.4	3.7
Railroads	5.2	5.8
(Nonrail)	2.1	2.3
Local railroads and bus lines		-1.0
Intercity passenger		1.5
Intercity motor trucking		3.1
Water transportation	0.5	0.7
Air transportation	8.0	8.2
Pipelines		9.1
Communications and public utilities	4.0	5.8
Communications	3.8	5.5
Electric, gas and sanitary service	3.9	6.1
Electric and gas	4.9	7.1
Man-made	2.5	2.9
Wholesale	2.5	3.1
Retail	2.4	2.7
Finance, insurance and real estate	NA	2.1
Services (except households; incl. Government enterprises)	NA	1.2

Source: Kendrick (1973), from tables 5.1 and 5.5.

The consequences of the imbalance of rates of productivity growth have been profound. We have experienced sharply rising relative costs and prices in the slow productivity growth sectors, and it is apparent that there is a widely held discontent with the performance of industries like housing, the complex of service industries, and with governmentally provided services like garbage collection and street cleaning.[2] On the other hand, for many of the goods and services produced by the rapid productivity growth sectors, there is an apparent sense of surfeit. There is at once a reaction of déjà vu regarding the coming generation of supersonic commercial aircraft, and a strong felt need to improve the railroads.

Second, to the extent that the argument above is accepted, the hunt for appropriate policy instruments will not be an easy one. Macro measures will not do; thus proposals like a general R&D tax credit (which has been quite fashionable in the recent discussions in the US) are beside the point. Policies need to be designed to influence particular economic sectors and activities. Regarding these, the key policy problem will be to augment or redesign institutions rather than to achieve particular resource allocations per se. Improving the railroads does not look like an objective that can be met through funding a few well specified R&D projects. Rather, the policy search must be for a set of institutions that will allocate resources appropriately over a wide range of circumstances and time.

Third, the character of the appropriate institutional structure for the generation, screening, and effective exploitation of innovation depends on the underlying technologies, the nature of the demands for the goods and services, and the characteristics of the organizations supplying them. These critical variables differ from sector to sector. General analytic arguments, for example about externalities that are inherent in certain kinds of R&D, have little bite since they ignore sectoral differences. Useful analysis must focus on and illuminate these differences.

While these arguments will be developed in more detail shortly, we believe to some extent the plausibility of these basic premises is apparent on their face. If they are accepted then, to be useful to thinking about policy, theory must be wide enough to relate the technological progressivity of the sector to the institutional structure involved. And the theory must be strong enough to guide plausible thinking about the effects of the various changes in institutional structure. It is apparent that any useful and coherent theory of innovation must recognize explicitly the factors that differ across industries. Our present bag of theory of innovation does not do this, and thus is not very useful in the current policy dialogue.

Section 2 will review selected aspects of the literature bearing on the interindustry productivity puzzle, attempting to pinpoint where the real weaknesses are, and to identify the essential building blocks for a mere useful theory of innovation. Sections 3 and 4 present a way of theorizing about innovation that we believe has promise of broadening and strengthening understanding of the innovation process and the variables impinging on it, so as to improve our ability to design efficacious policy. We certainly do not profess to have yet achieved a complete intellectual edifice. However, we can show you the floor plans, and point to certain parts of the structure that are taking shape here and there. In section 5 we shall sketch a few questions relating to institutional structure.

2. The State of Current Understanding

The weakness of present understanding of the reasons behind the differential productivity growth puzzle is in part due to lack of facts. But it is due at least as much to lack of theory that will enable us to knit together and give structure to what we know and extend our knowledge beyond particular facts. While there has been a considerable volume of research by economists, other social scientists, and historians of science and technology, that ought to bear on the differential productivity puzzle, that research is not well connected. This makes review, much less integration, of what is known quite difficult. More important, it means that knowledge is in the form of congeries of semi-isolated facts, rather than a connected intellectual structure.

2.1 The economists' model of differential productivity growth

There has been, however, one cluster of research that has been aimed directly at the differential productivity growth puzzle, and which has had a sufficiently strong theoretical structure so that knowledge is relatively integrated and has been cumulative. This structure derives of the text book economic theory of production. Studies within this tradition generally take the form of regression analyses attempting to explain differential productivity growth by differences in research and development and other factors which are seen as 'shifting' the production function.

Kendrick's early study (1973), particularly the work of Terleckyj contained in that study, set the style for much of the subsequent work based on production theory. The analysis was concerned with explaining cross sectoral differences in growth of total factor productivity, rather than output per worker. While the two measures are highly correlated, it is significant that the analysis simply took as given productivity growth accounted for by increases in capital per worker, and aimed to explain the 'residual'. For our purposes, the primary conclusion was that research and development intensity of an industry (measured in several different ways) was a significant factor explaining differences in total factor productivity growth across manufacturing industries (the only industries for which research and development data existed) both in simple and multiple regressions. Several other industry characteristics, for example growth of total output in the industry and sensitivity of output to cyclical fluctuations, also were strongly correlated with growth of total factor productivity. For these variables the possibility of two-way causation was recognized explicitly. For the most part, subsequent studies attempting to explain across industry differences in rates of productivity growth have followed roughly in this pattern set by Kendrick—Terleckyj, but with a wider and finer scan of variables and industries, and (in some cases) more concern with theoretical specification of the regression equations.

Mansfield's study (1968), as Kendrick—Terleckyj, was limited to a sample of manufacturing industries and was focused in the relationship between growth of total factor productivity and R&D spending. Unlike Kendrick—Terleckyj, Mansfield worked with an explicit production function formulation in which accumulated and depreciated research and development expenditures were treated as a form of capital. A number of different specifications were explored; for example, both embodied and disembodied technical progress models were regressed. Like Kendrick—Terleckyj, Mansfield found a significant effect of an industry's research and development spending upon its measured rate of technical progress. This conclusion was robust under different specifications, although the magnitude was sensitive to the exact specification.

Leonard's (1971) study also dealt only with manufacturing, His most significant contribution to the analytic dialogue was a separation of research and development spending financed by the industry itself, and research and development spending done in an industry but Financed by government. His empirical conclusion was that self-financed research and development spending had a much larger impact than governmentally financed research and development upon both growth of industry output, and growth of output per worker. Leonard did not work with a total factor productivity measure, although it is apparent that he would have got the same qualitative results using it as he got with his output per worker measure. It is an open question as to whether his results are identifying a real difference between the effectiveness of funds from different sources that would apply across the industry spectrum, or whether his procedure simply has separated out a few major industries, like aircraft and missiles, and electronics, where the government is both a significant buyer, and a significant research and development financer. If the latter, government research and development spending may be a proxy for several special factors that obtain in these industries. One might conjecture that in the defense industries R&D has been pushed into a regime of very low marginal returns, and that this is what the low weight on government R&D is capturing. Also, there are difficulties (which may exert a downward bias to measured technical advance) with the price deflator used to measure growth of real output in industries (like the aerospace complex) which sell specially designed new products to a monopolistic buyer.

Brown and Conrad developed a quite complex theoretical specification of the factors influencing the rate of growth of industry output, which, however, was simplified significantly as the authors moved from the theoretical equations to the regression equations actually fitted (Brown and Conrad, 1967). Their sample of industries, as those in the studies listed above, included only those in the manufacturing sector. Their major contribution to the analytical dialogue was to include in their regressions a measure of research and development done by other industries and embodied in the intermediate goods purchased by the industry in question, as well as research and development funding of the particular industry. Their regression results show a significant impact of such indirect research and development. In Brown and Conrad, own and indirect R&D is simply added together. In a later study, Raines entered these separately in his regressions (Raines, no date). His results indicated, strangely enough, that indirect research and development spending seems to have a more powerful effect than own research and development spending.

Perhaps the most interesting recent study is that by Terleckyj (1974). He considers nonmanufacturing industries as well as manufacturing industries. This is an important and provocative extension. The average rate of total factor productivity growth in the nonmanufacturing sector has been at least as fast as that in the manufacturing sector, and yet their own research and development intensity is drastically lower in nonmanufacturing than in manufacturing. In his analysis of the effect of own research and development spending, Terleckyj found that applied research and development funding provided by companies in an industry was a more powerful explanation of differences in productivity growth across manufacturing industries than total research and development expenditures done by the industry. Since the principle difference between the two measures is government financed research and development, his findings are consistent with those of Leonard.

Terleckyj's most significant contribution is an enriched and sophisticated extension of the earlier work of Brown and Conrad, and Raines, assessing the contribution of research and development done by other industries and embodied in inputs. He distinguishes between research and development embodied in capital equipment, and in intermediate inputs. In simple regressions, both are significant factors in explaining differences in productivity growth across the spectrum of industries, nonmanufacturing as well as manufacturing. However, when own R&D and R&D embodied in inputs are both placed in multiple regression equations, own research and development spending does not carry much explanatory power although it is slightly more important for manufacturing than for nonmanufacturing. This conclusion, anticipated by Raines, could result from a variety of factors. One possibility is that research and development spending financed by an industry is largely focused on new product design that, because of weaknesses in price indices, does not show up adequately in measures of growth of industry output; in contrast process improvements, which show up more reliably in increased productivity, stem largely from improved inputs and capital equipment. Another possibility is that own R&D spending and purchased inputs from R&D intensive industries are strong complements. This would show up if the regression were correctly specified. But under slight misspecification the regression weight can shift to one variable or the other.

The studies have been useful and provocative, but have not cut very deep. Severe measurement problems make interpretation of some of the statistical results, difficult. Some of these measurement problems were sketched above. Griliches' recent work (1973) provides a wider catalogue.

But the fundamental problem is that of specification. The regression equations involve complex interactions in which factors influencing the demand side and the supply side of the technical change process are intertwined, and confounded with other influences. Kendrick's (1973) recent analysis shows this sharply. In simple co-relational analysis, he finds that productivity growth in an industry is related to research and development spending. However, productivity growth also is strongly correlated with growth of the output of the industry over time. When Kendrick employs a step-wise multiple regression technique, the two explanatory variables that carry the most weight in explaining productivity growth are growth of output, and extent of unionization (which has a negative effect). With these variables included, other variables like research and development spending add little to the goodness of fit. But interpretation is hazardous. Obviously, there are a tangle of causations, from R&D to productivity growth, from productivity growth and lowered prices to growth of output, from growth of output in the presence of scale economies to productivity growth, from expansion of the industry to greater incentives for R&D, and so on. One also can wonder whether unionization is a cause, or an effect, of sluggish technical change in an industry. While Kendrick's particular analytic strategy makes this problem explicit, it is implicit in the studies above which simply related productivity growth to research and development spending.

And even were it granted that the causation runs, at least in part, from research and development spending to industry progressivity, what explains the uneven allocation of research and development spending? Among the range of possible explanations two stand in stark contrast regarding their policy implications. One is that research and development activity is more powerful when directed toward the technologies of certain industries than toward the technologies of others; therefore, the disparities in rates of technical progress reflect some kind of innate differences on ability to advance efficiently the different kinds of technologies. The second possible explanation (not mutually exclusive) focuses not on possible innate differences in what R&D can do in different sectors, but on differences in institutional structure that influence the extent to which R&D spending is optimal and the results of R&D effectively employed. The proposition is that industries differ significantly in the extent to which the results of research and development spending are internalized by the sponsoring firms, that in some industries but not in others there is significant-government subsidization of research and development where external-ities are important, and that industries also differ significantly in the speed and reliability of the mechanisms by which new technology is screened, and the use of efficacious innovation spread throughout the sector. Needless to say, the differences in these explanations matter profoundly in terms of their policy implications.

Some headway can be made on the key questions by augmenting and enriching the basic production function framework above. For example, R&D spending might be treated *not* as an independent variable but as a variable to be explained by other factors, some institutional, some proxies for possible innate differences across industries. While, to our knowledge, this kind of integrated analysis is rare, the 'productivity growth as a result of R&D' studies are quite conformable with other studies relating R&D spending in an industry to firm size and concentra-tion variables. Similarly, it ought to be possible to try to explain the magnitude and perhaps the productivity of applied research and development spending in an industry by certain proxy indices of the strength of the scientific base.

We would argue, however, that the breadth and strength of the production function framework is inherently limited. To obtain a more solid understanding of innovation and what can be done to influence innovation, it is necessary to study in some considerable detail the processes involved and the way in which institutions support and mold these processes. Since the 'production function framework' contains at best a rudimentary characterization of process and relevant institutional structure, a considerably more fine grained theoretical structure is needed for these more microscopic studies.

2.2 Building blocks for a broader theoretical structure

A considerable body of research has attempted to take a more microscopic look.[3] There is a rapidly increasing literature on the nature of the research and development process, the links between science and invention, the sources of invention (large firms, small firms, private inventors), the kinds of organizational and other factors associated with successful choice and carrying out of a project, etc. Other studies have probed at learning phenomena and, more generally, the way technologies (or a particular technology) evolve over time. A significant literature exists on organizational factors influencing the decision to adopt an innovation. Diffusion of innovation has been a fertile research field in several disciplines. While, for the most part these studies have not been concerned explicitly with inter-industry differences, innovations in many different industries have been studied, and it ought to be possible to make some cross-industry comparisons.

Unfortunately, these studies add much less than we would hope to our understanding of the differential productivity growth puzzle. The basic problem was stated briefly earlier – by and large, these studies have proceeded within disjoint theoretical frameworks. These are virtually no conceptual bridges between project SAPPHO which probes at conditions for successful innovation (SPRU, 1972), the Jewkes, Sawers, and Stillerman (1961) study of the sources of invention, and studies by economists such as Mansfield (1968) and Griliches (1957) on diffusion. Most important, there is no way to link these studies and the body of research concerned directly with differential productivity growth. Our knowledge is Balkanized. We cannot, in general, bring together several different bodies of analysis to focus on any one question, or tie together the various pieces to achieve an integrated broader perspective. Thus, while a considerable amount of research had probed at the details of process, and at the nature of prevailing institutional structure and how this has influenced process, there is no way in which we can confidentially link this knowledge to our understanding of the factors behind the differential productivity growth rates we have experienced. However, the micro studies have documented a number of 'facts' about innovation with which any widely embracing theory must be consistent. Two of these facts indicate that it is not promising to use the theoretical structure behind the productivity growth studies as a starting point.

The first of these facts is that innovation involves uncertainty in an essential way. The implicit process characterization of the 'production function' models would appear to be not only rudimentary, but fundamentally misleading. The problem cannot be patched up by reposing the theory in terms of expectations, introducing considerations of risk aversion, etc. Rather, a theoretical structure must encompass an essential diversity and disequilibrium of choices. Because of the uncertainty involved, different people, and different organizations, will disagree as to where to place their R&D chips, and on when to make their bets. Some will be proved right and some wrong. Explicit recognition of uncertainty is important in thinking about policy. One fundamental implication is that it is desirable for the institutional structure to generate a variety of innovations. Another implication is that a major function of an effective institutional structure is that it screen innovations effectively, accepting and spreading the good, winnowing out the bad.

A second fact that the microcosmic studies have illuminated is that the institutional structure for innovation often is quite complex within an economic sector, and varies significantly between economic sectors. Thus, in agriculture, there is considerable public subsidization of research done by predominantly non-profit institutions (largely universities) and a subsidized federal-state extension service for the dissemination of information regarding new technological developments to farmers, interacting with the network of private farms, and industries that produce and sell farm equipment, fertilizers, etc. The commercial aircraft industry is equally complex, but must be described in quite different terms. Innovation in medicine involves a set of institutions different from either of these. This institutional complexity and diversity would seem to be where the focus of a policy attention should be; however, it does not seem possible to extend the rudimentary institutional assumptions built into the production function model sufficiently to really grip these dimensions.

If there is to be any hope of integrating the disparate pieces of knowledge about the innovation process, a theory of innovation must incorporate explicitly the stochastic evolutionary nature of innovation, and must have considerable room for organizational complexity and diversity. Merely having these attributes is, of course, no guarantee that a theoretical structure will be able to integrate what is known, or have the power to predict the effect of reasonable changes in institutional structure. But, these attributes can give some guidance to the search for a useful theory.

Over the past several years, we have been developing the outlines of such a theoretical structure.[4] Our objective is to develop a class of models based on the following premises. First, in contrast with the production function oriented studies discussed earlier, we posit that almost any nontrivial change in product or process, if there has been no prior experience, is an innovation. That is, we abandon the sharp distinction between moving along a production function and shift to a new one that characterizes the studies surveyed earlier. Second, we treat any innovation as involving considerable uncertainty both before it is ready for introduction to the economy, and even after it is introduced, and thus we view the innovation process as involving a continuing disequilibrium. At any time there is coexistence of ideas that will evolve into successful innovations and those that will not, and actual use of misjudged or obsolete technologies along with profitable ones. Over time selection operates on the existing set of technologies, but new ones continually are introduced to upset the movement toward equilibrium.

More formally, in an accounting sense we view productivity growth as explained within our proposed theoretical structure in terms of first, the generation of new technologies, and second, changes in the weights associated with the use of existing technologies. This accounting distinction reflects an analytic break. We are attempting to build conformable sub-theories of the processes that lead up to a new technology ready for trial use, and of what we call the selection environment that takes the flow of innovations as given, (Of course, there are important feedbacks.)

The structure we propose is capable of quite formal articulation and modeling. We have built several formal models and reported on these in earlier papers. In this paper, the emphasis will be on the theory as providing a useful language for talking about dynamic process, for integrating presently disparate knowledge, and for guiding future research.

3. The Generation of Innovation

Fragmentation of empirical work certainly characterizes the state of knowledge regarding the genesis of innovation. However, almost all students of innovation agree that purposive acts of investment are an important part of the process. Many scholars have recognized two classes of factors influencing the allocation of effort: factors that influence the demand for or pay-off from innovation, and factors that influence the difficulty or cost of innovation.

Demand side factors have been studied by a variety of research traditions. In economics, Schmookler (1966), for example, has articulated and provided strong supporting evidence for a simple model in which changes in the composition of demand for goods and services across industries chain back to influence investment patterns, which in turn influence the relative return to inventors working on improvements in different kinds of machines. The theoretical work by economists on factor bias of innovation, and the empirical work which started with Habakkuk (1962), also is focused on the effect of demand side factors. Recently Hayami, Ruttan and followers, have provided strong empirical support for the proposition that relative factor prices influence the nature of innovation, at least in agriculture.[5] A wide variety of more micro studies, by economists and other social scientists, examining particular inventions, or the influences that bear on allocation decisions within a particular R&D organization, also have identified the importance of perception of demand or pay off.

Other studies have focused on the cost or supply side. In contrast with research on demand side factors, research aimed at exploring differences in the, difficulty or cost of different kinds of innovation has had but limited conceptual and empirical pay-off, A list of variables and conjectures has been proposed, for example, that the scientific base for innovation differs from industry to industry, or that innovation is intrinsically easier in industries that produce physical products than in those that produce services. It also has been proposed that mechanization, where relevant, is a particularly easy route for innovation and similarly, that latent scale economies often provide a route that is easy to follow.[6] Empirical support for these propositions, however, has been weak. Nonetheless, there would appear to be only a few dissenters to the proposition that factors on the cost side as well as factors on the demand side differ across industries and technologies, and that these differences are important in explaining the pattern of innovation that has occurred.

3.1 The profit maximization hypothesis and its limitations

The theoretical problem is how to organize what we know so that the whole adds up to more than the sum of the parts and knowledge extends beyond the particulars. At first thought it is tempting to adopt the economists' model of the profit maximizing firm as the basis for a theory of innovation genesis. The semi-empirical fact of purposeful and intelligent behavior regarding the allocation of resources to innovation, which almost all scholars believe characterizes the true situation, would be theoretically translated as profit maximizing behavior. Factors on the demand and supply side then would be brought in as they are in the traditional theory of the firm.

The models aiming to explain differential productivity growth, discussed in the previous section, in general have taken R&D expenditures as given, rather than as something to be explained. However, it seems apparent that the authors had in mind something like a profit maximizing model behind the scenes. And several profit maximizing models aimed at exploring R&D spending have been constructed. For example, it is possible to deduce the Schmookler conclusions simply by treating accumulated R&D spending as a form of capital, and extending the traditional result in the theory of the firm regarding the affects of an increase in demand for a product upon the amount of investment it is profitable to undertake. The literature by economists on factor bias, mentioned above, is self consciously within the 'profit maximizing' frame. The key concept is that of an innovation possibility set associated with a given level of expenditure, or with different elements associated with costs of different amounts. Firms are assumed to choose the profit maximizing element.[7] The effects of changes in factor prices on the direction of inventing can be deduced within this model.

A profit maximizing interpretation of purposive and intelligent behavior clearly has considerable organizing power. However, there are certain important difficulties with this formulation that need to be recognized.

The basic limitations relate to our remarks in the preceding section. In many cases, the organizations doing R&D are not motivated by profits at all, but are governmental, or private not-for-profit institutions. The difficulty here can be resolved on the surface by treating the term 'profit' very broadly to stand for whatever objectives the organizations happen to have. We shall adopt such a tactic in the following section. But, at the least, this calls for rather detailed empirical inquiry as to the nature of the organizations doing the R&D before the 'profit' concept can be made operational in a model. And, actually, the problem is even more complicated than that. In many sectors there are a complex of R&D organizations, some profit oriented, some governmental, some academic, doing different things, but interacting in a synergistic way. In particular, in medicine, agriculture, and several other sectors, private for-profit organizations do the bulk of R&D that leads to marketable products, but academic institutions play a major role in creating basic knowledge and data used in the more applied work.

Relatedly, everyone agrees that R&D is an uncertain business. Uncertainty resides at the level of the individual project, where the 'best' way to proceed seldom is apparent and the individuals involved instead have to be satisfied with finding a promising way. Uncertainty also resides at the level of R&D project selection. The enormousness of the set of possible projects, the inability to make quick cheap reliable estimates of benefits and costs, and the lack of convenient topological properties to permit sequential search to home in rapidly on good projects independently of where that search starts, means that project choice, as well as outcome given choice, must be treated as stochastic. The problem with the maximization metaphor is not that it connotes purpose and intelligence, but that it also connotes sharp and objective definition of the range of alternatives confronted and knowledge about their properties. Hence it suggests an unrealistic degree of inevitability and correctness in the choices made, represses the fact that interpersonal and interorganizational differences in judgment and perception matter a lot, and that it is not all clear ex-ante, except perhaps to God, what is the right thing to do.

Largely because of these limitations, the profit maximization formulation has been unable to cope with certain salient aspects of the innovation generation process. While it has had qualitative success in dealing with certain variables on the demand side, it has tended to ignore externalities, implicitly presuming that 'consumer' valuations are reflected in profit opportunities for firms doing R&D. It also has neglected the range of questions relating to how consumers find out about and value new products and processes. On both of these matters, the particular structure of institutions, and the way uncertainties are resolved, are key. Our discussion of the 'selection environment' for innovations is aimed to deal with these kinds of problems more coherently.

The profit maximization formulation has had very little grip on factors on the cost or feasibility of innovation side. Here a more process oriented characterization of the innovation generation process seems essential. We shall sketch such a characterization in the remainder of this section.

3.2. R&D strategies and probabilistic outcomes

In this section we outline what we believe may be a fruitful way to deal with some of these difficulties. The components of our proposal are these. An R&D project, and the procedures used by an R&D organization to identify and screen R&D projects, can be viewed as interacting heuristic search processes. A quasi stable commitment to a particular set of heuristics regarding R & D project selection can be regarded as an R&D strategy.[8] Often it is possible to identify a few R&D strategies that are prevalent in a particular sector in a particular era. An R&D strategy might be modeled extensively in terms of the heuristics employed in the search processes and their consequences. Or a strategy can be represented in terms of a conditional probability distribution of innovations (or innovation characteristics) given certain conditions facing the organization.

By a heuristic search process we mean an activity that has a goal, and a set of procedures for identifying, screening, and homing in on promising ways to get to that objective or close to it.[9] The procedures may be characterized in terms of the employment of proximate targets, special attention to certain cues and clues, and various rules of thumb. While they may be fruitful in the sense of yielding relatively satisfactory outcomes a good percentage of the time, they do not guarantee a good outcome or even a unique one. That is, they are heuristics, rather than an algorithm for calculating an optimum. While a costless maximizing algorithm would be preferred by decision makers to good heuristics, in complex decision problems a maximization algorithm is likely to be highly expensive to employ, even if one is known. In most cases one simply doesn't exist. Thus good heuristics is the best one can hope for.

The notion that an R&D project can be viewed as a search is quite widespread in the literature. Recently, Evanson and Kislev (1975) have developed a specific model of the search process involved in trying to find a new seed variety with improved performance in certain dimensions. They place their model in a maximizing framework, in the sense that they calculate the number of elements (seed varieties) that 'should' be sampled. However, their model glides over exactly those aspects of an R&D project where heuristics are important – the judgments as to what elements to sample first, assessment of what has been learned from the early draws that provides guidance as to the next steps, etc. We shall not elaborate here the notion of intra project heuristics save for one point. We. propose that an important aspect of the question of the role of science in invention can fruitfully be posed in terms of the ways in which scientific knowledge enables powerful search heuristics.

While this has been less extensively discussed, it is apparent that the project selection process also must be heuristic. For all of the reasons discussed above, the selection by a firm of a set of projects to pursue cannot be maximizing in any serious sense. Rather, the process must employ various proximate targets, attend to particular cues and clues, use various rules of thumb.

One of the characteristics of heuristics is that often they factor a complex interrelated decision problem into parts, which then are treated as more or less independent. For example, obviously decisions regarding the size of the overall R&D budget, and decisions regarding the projects that will be undertaken, are strongly interdependent. However, there is considerable evidence that many business firms employ R&D decision heuristics that separates these dimensions, at least at first cut. Thus setting the R&D budget as a certain fraction of sales seems to be a widely used heuristic for the first stage, at least, of the budgetary process (Mansfield, 1968).

There similarly seem to be certain widely used heuristics for hunting for promising projects to undertake, which proceed independently of the research budget, save for an 'adding up' constraint.

It is convenient to employ the term 'strategy' to refer to a stable widely used set of heuristics for project selection. A model of a strategy ideally would be able to pick out, probabilistically, the nature of the R&D projects that would be selected by an organization employing it, given certain conditions under which project selection proceeds, for example, the pattern of consumer demand for different products and product improvements, and the state of scientific knowledge (which affect project ranking) and total firm sales (which affect the firm's total R&D budget). The outcomes, in terms of the degree and kind of success achieved from the projects actually undertaken, also will be probabilistic, given project heuristics and external conditions. Since project selection implies project heuristics, an R&D strategy can be viewed as defining a probability distribution of number and kinds of innovations, given certain variables that influence project selection and project outcome.

A considerable portion of research on the generation of innovation can be viewed as attempts to find, describe, and analyze the consequences of the employment of certain widely studied strategies. One would expect that the nature of strategies would differ greatly depending on the nature of the organization being studied. For example, government agencies would have different strategies than business firms; however, for the most part research has focused on business firms, and this will be our focus here. A good R&D strategy must attend to factors on the demand side and factors on the supply side. It is no good to pick out projects that are technologically exciting and doable, but which have no demand, or to undertake projects which if successful would have a high payoff, but where there is no chance of success. However, one can imagine a strategy that focuses on the 'demand' side and picks out a list of inventions that if made would have a good market, for later screening according to feasibility and cost. Or one can imagine a strategy that initially picks out projects where major technological breakthroughs seem possible, for later checks for market ability. A considerable literature has evolved that implicitly posits that R&D-strategies can be dichotomized into these two. The first strategy has been named 'demand-pull'; the second 'capabilities-push'.[10] Both, presumably, pick out projects within some kind of a prior budget constraint (say R&D as a certain percent of sales).

Of course, if one were committed to the proposition that project choice was truly maximizing, then it should make no difference whether choice proceeded by first listing projects with good demand prospects and then doing a cost or feasibility check on this list, or by prescreening of technical feasibility and then checking for demand. The order of calculation would be irrelevant. However, if one views the first cut as defining a region for consideration, with subsequent more detailed calculation over the selected subset, demand-pull and capabilities-push are quite different strategies. In a given external regime of stimulus, they will select different projects, and presumably they will have different consequences in terms of the payoffs achieved.

Several different studies have concluded that, if strategies can be so dichotomized, demand-pull is by far the more common of the two (Pavitt, 1971). Further, when applied, demand-pull is more likely to result in a commercially successful project than a strategy of capabilities-push. However, capabilities-push selected projects, when they do pay off, pay off handsomely. None of these results is surprising. One might well have expected that screening first for innovations which, if they could be achieved, would yield high payoffs, would focus attention on a more fruitful part of the choice set than screening first for things that techno-logical wizardly can accomplish. One also might have expected that, when the R&D cost or feasibility criteria were applied to a project list preselected on demand considerations, the outcome would be a project the objectives of which can be achieved with modest cost and high confidence. However, the 'look first at technological possibilities' strategy should occasionally come up with a big winner. Thus categorizing R&D strategies in this way would appear to be a fruitful intellectual endeavor.

However, if one thinks about it for a moment, both a pure demand-pull and a pure capabilities-push strategy would appear to be naive. One might posit that a strategy that involves more backing and forthing between demand and supply side considerations would be more powerful. Further, it is patently implausible that project generators look first at 'all' R&D projects with high demands, or at 'all' major achievable breakthroughs. These sets are too vast. Also, the very staffing of an R&D organization inevitably limits the range of projects it has the capabilities to undertake.

We propose that most R&D organizations de facto are limited to strategies which involve a precommitment to one or a small number of classes of R&D projects each of which has a certain similarity of broadly defined targets, procedures for reaching these targets, and R&D resources required. Binswanger (1974) has used the term 'R&D activity' in a way that captures what we mean here. Following the work of Hayami and Ruttan, he notes that in agriculture one can identify various classes of R&D project. One can distinguish between projects aimed to improve mechanical equipment, projects that aim to improve seed varieties, etc. Within these classes both the targets, and the procedures followed by those doing the R&D, have a certain similarity. Relatedly, an R&D activity defines, or is defined by, certain kinds of skills, equipment, and organization. He proposes that most R&D projects, at least in agriculture, can be viewed as belonging to one of a relatively small group of R&D activities, in the sense above. An R&D organization can staff itself to work with one or a few of these R&D activities. This involves a strong limitation on its R&D strategy. Given this precommitment, demand-pull or technology-push may guide the period by period details of choice. Or other procedures may be employed.

In the Binswanger formulation the set of projects associated with a particular R&D activity is closely circumscribed. Thus there is no way to try to adjust for changes in demand conditions, or changing costs of R&D, operating within that activity. Rather, changes in the kind of R&D done, in response to changes in external conditions, must come about through changes in the mix of R&D activities employed. We see no inevitable reason why the examples of R&D activity, given by Binswanger, and Hayami and Ruttan, must possess this internal inflexibility. For example, mechanical devices comprise a large class of items, and the various members of the class each possess a number of relevant dimensions. Certainly, R&D aimed at improving mechanical design can aim in a variety of directions. Similarly, there are a number of possible attributes one might aim for in trying to develop a new seed variety. In any case, the economists' bias is that a viable strategy ought to be able to adjust to changing demand and cost conditions, and thus must involve a mix of activities, or a very flexible activity.

However, it may be that there are certain powerful intra project heuristics that apply when a technology is advanced in a certain direction, and payoffs from advancing in that direction that exist under a wide range of demand conditions. We call these directions 'natural trajectories'. If natural trajectories exist, following these may be a good strategy.

3.3 Natural trajectories

In some of the writing on technological advance, there is a sense that innovation has a certain inner logic of its own. In contrast with the central focus of the economists' models - a fine tuned responsiveness to demand conditions and their changes, and a scan of a wide range of projects to assess profitability - particularly in industries where technological advance is very rapid, advances seem to follow advances in a way that appears somewhat 'inevitable' and certainly not fine tuned to the changing demand and cost conditions. Rosenberg talks of 'technological imperatives' as guiding the evolution of certain technologies; bottlenecks in connected processes, obvious weak spots in products, clear targets for improvement, etc. (Rosenberg, 1969, 1972). These provide sharp signals that certain R&D projects are doable and are worth doing under a wide range of particular demand and cost conditions. Marginal changes in external conditions influence at most the ranking in terms of profitability of the set of profitable projects associated with pushing technology in a particular direction. We suggest that such natural trajectories are important, and can be studied.

In many cases natural trajectories are specific to a particular technology or broadly defined 'technological regime'. We use the 'technological regime' language to refer to much the same kind of thing as Hayami and Ruttan mean by a meta production function. Their concept refers to a frontier of achievable capabilities, defined in the relevant economic dimensions, limited by physical, biological, and other constraints, given a broadly defined way of doing things. Our concept is more cognitive, relating to technicians' beliefs about what is feasible or at least worth attempting. For example, the advent of the DC3 aircraft in the 1930's defined a particular techno-logical regime; metal skin, low wing, piston powered planes. Engineers had some strong notions regarding the potential of this regime. For more than two decades innovation in aircraft design essentially involved better exploitation of this potential; improving the engines, enlargening the planes, making them more efficient.[11]

Binswangers R&D activities, our R&D strategies, often are inti-mately connected with a given technological regime, in a sense not well developed by Binswanger. The sense of potential, of constraints, and of not yet exploited opportunities, implicit in a regime focuses the attention of engineers on certain directions in which progress is possible, and provides strong guidance as to the tactics likely to be fruitful for probing in that direction. In other words, a regime not only defines boundaries, but also trajectories to those boundaries. Indeed these concepts are integral, the boundaries being defined as the limits of following various design trajectories.

In many cases the promising trajectories and strategies for tech-nological advance, within a given regime, are associated with improve-ments of major components or aspects thereof. In aviation, engineers can work on improving the thrust-weight ratio of engines, or on increasing the lift-drag ratio of airframes. General theoretical understanding provides clues as to how to proceed. In jet engine technology, thermodynamic understanding relates the performance of the engine to such variables as temperature and pressure at combustion. This naturally leads designers to look for engine designs that will enable higher inlet temperatures, and higher pressures. In airframe design, theoretical understanding (at a relatively mundane level) always has indicated that there are advantages of getting a plane to fly higher where air resistance is lower. This leads designers to think of pressurizing the cabin, demand-ing aircraft engines that will operate effectively at higher altitudes, etc.

Often there are complementariness among the various trajectories. Advances in engine power, and streamlining of aircraft, are complementary. Developing of seeds that germinate at the same time and grow at the same rate facilitates mechanical harvesting.

While natural trajectories almost invariably have special elements associated with the particular technology in question, in any era there appear to be certain natural trajectories that are common to a wide range of technologies. Two of these have been relatively well identified in the literature: progressive exploitation of latent scale economies, and increasing mechanization of operations that have been done by hand.

There are a wide variety of industries and technologies where the advance of equipment involves the exploitation of latent scale economies. In chemical process industries, in power generation, and in other sectors where designing of equipment of larger capacity will permit output expansion without a proportional increase in capital or other costs, the objectives of cost reduction apparently lead designers to focus on making equipment larger. Hughes has documented the way in which designers of electric power equipment have aimed progressively to push forward the scale frontier (Hughes, 1971). Levin has provided a general theoretical discussion of the phenomenon and provided case studies of the process in operation in the manufacture of sulfuric acid, ethylene, ammonia, and petroleum refining (Levin, 1974). In the development of aircraft technology, designers long have understood that larger planes in principle could operate with lower costs per seat mile. Of course in aviation, as in electric power, the possibilities for exploiting latent scale economies are limited by the market as well as by engineering. In aviation, high volumes and long hauls provide the market targets of opportunity. And, historically, these have tended to grow in importance over time. This has permitted engineers to follow their design instincts. As a rule, each generation of commercial aircraft has tended to involve larger vehicles than those in the predecessor generation. Another quite common natural trajectory is toward mechanization of processes that have been done by hand. Mechanization seems to be viewed by designers of equipment as a natural way to reduce costs, increase reliability and precision of production, gain more reliable control over operations, etc. This point has been stressed by Rosenberg in his study of 19th century innovation in American industry (Rosenberg, 1972). That this tendency to mechanize still exists has been suggested by Piore (1968), and documented in considerable detail by Setzer (1974) in her work on the evolution of production processes at Western Electric. Inventors and research and development engineers, operating under a higher order objective to look for inventions and design chances that will reduce costs, look for opportunities to mechanize. Engineers, through training and experience, apparently acquire heuristics that assist the design of machinery. For this reason, hunting for opportunities for mechanization, like trying to exploit latent scale economies, can serve as a useful focuser for inventive activity.

David (1974), in a fascinating and important recent essay, proposes a different but complementary hypothesis. While the studies above point to 'easy invention' in directions that increase the capital labor ratio, David suggests that in the late 19th century technologies that already were capital intensive were easier to improve in a 'neutral' direction, than were technologies that involved a lower degree of capital intensity. Here the proposition was that during that period of time there was a 'lot of room' for improving mechanized operations, and engineer-designers had some clever ways of moving in that direction.

Exploitation of latent scale economies, and opportunities for further mechanization, are important avenues for technological advance in many sectors at the present time, as well as in the 19th century. Many of the studies cited above are of relatively contemporary examples. However, there is no reason to believe, and many reasons to doubt, that the powerful general trajectories of one era are the powerful ones of the next. For example, it seems apparent that in the 20th century two widely used natural trajectories opened up (and later variegated) that were not available earlier: the exploitation of understanding of electricity and the resulting creation and improvement of electrical and later electronic components, and similar developments regarding chemical technologies. As with the case of mechanization during the 19th century, these developments had several different effects. For example, improvements in ability to understand electrical phenomena and growing experience with electrical and electronic equipment led to a substitution of these kinds of components for others. And technologies that had many and important electronic components were better able to benefit from the improvements in these components than other technologies.

It is apparent that industries differ significantly in the extent to which they can exploit the prevailing general natural trajectories, and that these differences influence the rise and fall of different industries and technologies. During the 19th century, cotton gained ascendency over wool in large part because its production processes were easier to mechanize. Quite possibly both Rosenberg trajectories and David trajectories were involved. In the 20th century, Texas cotton drove out southeastern cotton in good part because the area was amenable to mechanized picking. In the current era, where considerable power has been developed to design and improve synthetic products, synthetic fibers have risen in importance relative to natural ones.

One aspect of natural trajectories, whether specific to a particular technology or more general, whether 19th century or contemporary, is that underlying the movement along them is a certain knowledge on the part of the technicians, engineers, scientists, involved in the relevant inventive activity. The knowledge may be quite specific, as understanding of the tactics for hybrid development of seeds, or the operating character-istics of jet engines. The knowledge may involve more art and feel than science; this certainly was so of the knowledge behind the mechaniza-tion and scale economies of trajectories during the 19th century. But in the middle to late 20th century, many scholars have been strongly tempted by the hypothesis that underlying the technologies that have experienced the most rapid advance, or built into a key component of these, is a relatively well articulated scientific knowledge. This does not mean that the 'inventors' are active scientists, nor that 'inventing' exploits knowledge produced by recent science. But the fact that college educated scientists and engineers now comprise the dominant group doing applied research and development indicates that, at the least, scientific literacy is in the background. One then is led to the proposi-tion that a key factor that differs across industries and that partly resolves the differential productivity puzzle is the strength of scientific under-standing relevant to seeking improvements, We shall not review that tangled literature here. However, in the concluding section we will pose the question as to whether such differences, if important, are inherent or institutional.

Whether a 'strategy' involves a natural trajectory or not, whether or not there are certain identifiable R&D activities among which firms must choose some small subset, whatever the ways used to assess chances of benefits and feasibility, an R&D strategy determines R&D outcomes, probabilistically. There will be some winners and some losers. The next analytic question is, 'what next?'

4. The Selection Environment[12]

4.1 Elements of the selection model

The heart of the conceptualization problem discussed in the pre-ceding section was to characterize the generation of innovation as purposive, but inherently stochastic. Despite a tendency of some authors to try to slice neatly between invention, and adoption, with all of the uncertainty piled on the former, one cannot make sense of the micro studies of innovation unless one recognizes explicitly that many uncertainties cannot be resolved until an innovation actually has been tried in practice. While the organizations watching the flow of new innovations may be trying to behave rationally, as with the analysis of the generation of innovation, rational behavior should not be translated as maximizing behavior unless explicit recognition is made of differences in perception or luck. Relatedly, analysis of the ways that innovations are screened, some tried and rejected, others accepted and spread, must be explicitly dynamic. We propose the concept of a 'selection environment' as a useful theoretical organizer. Given a flow of new innovations, the selection environment (as we are employing the terms) determines how relative use of different technologies changes over time. The selection environment influences the path of productivity growth generated by any given innovation, and also it feeds back the influence strongly of the kinds of R&D that firms and industry will find profitable to undertake.

There is an important conceptual issue that needs to be clarified at the outset. In much of the literature on technological change a sharp distinction has been drawn between inventing and innovating (where the latter term is used, more narrowly than we are using it, to refer to a decision to try out technology in practice). The distinction harks back to Schumpeter of the *Theory of Economic Development* (1934). While technological invention was not center piece in his analysis, regarding invention he described a world in which independent inventors had to link up with firms in being, or entrepreneurs seeking to establish new firms, to implement their inventions. We would argue that in the current institutional environment with much of innovation coming from internal R&D, the old Schumpeterian distinction is much less useful than it used to be. While there are examples of inventions that were economically viable without further R&D that simply lay around waiting for some to try them out, this seems a rare occurrence. Further, the earlier experimental use of a new technology often is integrated with the last stages of the research and development process.

There is, however, a distinction that has some of the flavor of the old Schumpeterian distinction that ought to be recognized. Often an innovation is produced by a firm for sale to customers who will use it. Thus there are two acts of innovation (in the narrow sense of the term) that are involved. In the case of the advent of jet passenger aircraft, DeHaviland, the company that produced the first commercial jet, was an innovator. But so was the airline that bought the plane. More generally, if the focus is on any economic sector, it is useful to distinguish between two kinds of innovation. Some of these may bubble out of the research and development activities of the firms in the sector. Others may be largely in the form of materials, components or equipment offered by supplying firms. However, for the moment let us repress that distinction and focus on an economic sector which is experiencing a flow of new innovations, some of which may be viable, and others not. While the range of possible innovations, and the characteristics of the sectors, obviously are extremely diverse, the analytic task is to develop a conceptual framework which at once identifies commonalities and enables the differences to stand out.

Consider, then, the following diverse set of innovations and industries: the first model 707 aircraft produced by the Boeing Aircraft Company, the first use of the oxygen process on a commercial basis by a steel company in Austria, a new seed variety tried by a farmer, a pioneering doctor trying a anti-cancer drug, a district court trying the system of release on own recognizance without bail for select group of people accused of crime, a school trying an open classroom. The range of possible innovations, and the characteristics of the organizations that introduce them, is enormous.

A necessary condition for survival of an innovation is that, after a trial, it be perceived as worthwhile by the organizations that directly determine whether it is used or not. If the innovation is to persist and expand in use, the firm must find a new product or process profitable to produce or employ, the doctor must view the treatment as efficacious, the school system must be persuaded that the new classroom technique is good educational practice and worth the cost. We shall call all of these primary organizations 'firms' and use the term profitable to indicate value in the eyes of the firms, without implying that the objectives are money profit rather than something else, or that the organization is private, not public. Neither do we imply any social merit in firms' objectives. Firms may be motivated by little more than the prestige of being first. Sectors obviously differ in terms of the objectives of the firms.

The question of whether or not the firms find the innovations profitable depends not only on the objectives of the firms. In almost all economic sectors the firms – for-profit private organizations, public agencies, individual professionals – are subject to monitoring mechanisms which influence the innovations that score well or poorly according to the objectives of the firms and which may more directly impose constraints on firm behavior. A key part of this monitoring mechanism involves the individuals or organizations who are the demanders or beneficiaries of the goods or services produced by the firms in the sectors. Thus the profitability to Boeing of producing 707 type aircraft depends on how the airlines react to these planes. Consumers must be willing to buy the corn that the new seed produced at a price that covers cost. Patients must agree to the new treatment. School systems and legal systems are constrained by funds proposed by high order executives, and voted by legislatures. In some sectors there are additional constraints imposed on firms by agencies that are assigned a legal responsibility to monitor or regulate their activity. Thus the Boeing 707, before it could be put into commercial use, had to pass FAA tests. New Pharmaceuticals are regulated, etc. Selection environments differ greatly in the structure of demanders and monitors and the manner and strength in which these mold and constrain the behavior of firms.

There are, roughly speaking, two roughly distinct kinds of mechanisms for the spread of a profitable innovation. One of these is greater use of an innovation by the firm that first introduces it. If the firm produces a variety of products or undertakes a variety of activities, this may occur through substitution of the new activity for older ones. Or, the firm may grow, absolutely, and (if there are competitors) relatively by attracting new resources. In sectors which involve a number of administratively distinct organizational units on the supply side, there is a second innovation spreading mechanism that needs to be considered - imitation. Imitation of certain innovations may be deliberately spurred by the institutional machinery. Thus the agricultural extension service encourages widespread adoption by farmers of new seed varieties. If the innovation is produced by a supplying firm, its sales agents will try to encourage rapid adoption. Or the institutional machinery may deter or block imitation, as the patent system blocks the adoption by one firm of patented innovations created by a competitor.

The relative importance of these mechanisms differs from sector to sector. Dieselization of a nationalized railroad system must proceed largely through substitution of diesels for other kinds of railroad engines, although improvement in the service may enable a nationalized railroad system to gain additional funds. If, on the other hand, there are a number of organizationally separate railroad systems, when one railroad is a successful innovator, spread of diesels may to a considerable extent require imitation by other railroads. The success of the 707 encouraged and enabled Boeing to expand its production facilities. And other aircraft producers were spurred, at their peril, to design and produce comparable aircraft. Bail reform has spread in part by greater use within particular districts, but since one jurisdiction is not permitted to expand relative to another, and since there are many thousands of jurisdictional districts, the ultimate spread of innovations in the Criminal Justice System depends upon imitation.

We propose that a rigorous general model of the selection environment can be built from specification of these three elements: the definition of 'worth' or profit that is operative for the firms in the sector, the manner in which consumer and regulatory preferences and rules influence what is profitable, and the investment and imitation processes that are involved. In the remainder of this section we shall discuss some important qualitative differences in sectoral selection environments that become the focus of attention once one poses the theoretical problem in the way we have. Market sectors differ significantly among themselves. And many sectors involve important nonmarket components that have special characteristics.

4.2 The market as a selection environment

The perception that market competition in a sector operates like a selection environment was explicit in the writings of many of the great 19th and early 20th century economic theorists, Schumpeter was well within the classical tradition. In a stylized Schumpeterian evolutionary system, there is both a carrot and a stick to motivate firms to introduce 'better' production methods or products. Better here has an unambiguous meaning: lower cost of production, or a product that consumers are willing to buy at a price above cost. In either case the criterion boils down to higher monetary profit. Successful innovation leads to both higher profit for the innovator and to profitable investment opportunities. Thus profitable firms grow. In so doing they cut away the market for the noninnovators and reduce their profitability which, in turn, will force these firms to contract. Both the visible profits of the innovators and the losses experienced by the laggers stimulate the latter to try to imitate.

It would seem likely that the Schumpeterian dynamics would differ somewhat depending on whether the innovation were of a new product or a new process. For product innovation, the profitability to the firm is bound tightly to the reactions of potential consumers. For process innovation, which does not change the nature of the product, consumer constraints are far more blunt. The firm can make an assessment of profitability by considering the effects on costs, with far less concern for consumer reaction. Further, and reinforcing these differences, product innovation usually comes from a firm's own R&D; process innovations likely come from the R&D done by suppliers and are embodied in their products. To the extent this is so, imitation by a competitor of a process innovation is likely to occur relatively rapidly, and to be encouraged by a marketing supplier, rather than being retarded by a patent.

Both expansion of the innovator, and imitation by competitors are essential to the viability of Schumpeterian process. In the standard descriptions of dynamic competition, expansion of the innovator is likely to be stressed. It is surprising, therefore, that the relationship between innovation and investment has been studied hardly at all empirically. The principal studies of firm investment have been based on neoclassical theory modified by Keynesian considerations, and tend to ignore the relationship between innovation and expansion of a firm. The Meyer and Kuh (1957) retained earnings-capacity pressure theory would imply that successful innovators tend to expand . Presumably a successful innovation both yields profits, and attracts demand which may, initially at least, exceed capacity. A more straight laced neoclassical theory also would predict that firms that come up .with better processes and products ought to want to expand their capacity to produce. But the major studies of firm investment have, Virtually without exception, ignored the influence of innovation on investment.

The exceptions are studies where the author's basic hypothesis is oriented around the Schumpeterian interactions. Mueller does find that lagged R&D expenditure by a firm has a positive influence on its investment in new plant and equipment (Mueller, 1967), In a later study, Grabowski and Mueller used lagged patents as a measure of R&D output, but find that the influence on plant and equipment investment is weak statistically (Grabowski and Mueller, 1972). Mansfield's studies give stronger support for a 'Schumpeterian' view. In examining investment at an industry level, he finds that the number of recent innovations is a significant explanatory variable, augmenting more traditional variables (Mansfield, 1968). But perhaps his most interesting results involve comparisons of firm growth rates, where he finds that innovating firms in fact tend to grow more rapidly than the laggers. However, while the advantage of the innovators tends to persist for several periods, the advantage tends to damp out with time, apparently because other firms have been able to imitate, or to come up with comparable or superior innovations.

In contrast to the sparseness of studies of the relationship of investment to successful innovation, a large number of studies have focused on the spread of innovation by diffusion (imitation) in profit-oriented sectors.[13] There have ranged across a variety of sectors from agriculture (study of the diffusion of hybrid corn among farmers) to railroads (diesel engines) to brewing, to steel. Many have documented the role of profitability of an innovation in influencing the speed at which that innovation spreads. However, other studies have concluded that the calculations made by firms tend to be haphazard, and that even ex-post the firms had little idea, quantitatively, how profitable the innovation turned out to be (Nasbeth and Ray, 1974). Several have found that, for innovations that are costly to put into operation, large firms (with greater financial resources) tend to adopt a new technology earlier than do smaller firms, although there are exceptions. Most of the studies show an S shaped pattern of use of the new innovation over time. In many cases this has been attributed to the fact that the later users are observing the behavior (and perhaps performance) of the earlier adopters before making their own decisions. In some instances the innovations were inputs provided by a supplier, and the early adopters of the innovation were not in a position to block subsequent use of their competitors. In other instances this was not the case. For example, a glass producing company, Pilkington, holds the basic patents on the float glass process and presumably had an interest in limiting diffusion to other firms except where Pilkington was blocked from the market. It is interesting that the analysts of diffusion have not in general been cognizant of these differences.

It also is quite surprising that in no study of which we are aware has there been an attempt to study the dual roles of expansion of the innovator, and imitation of the imitator, together. It would seem apparent that in order for a market selection environment to work effectively, a rather fine balance is required between the two mechanisms. We will return to this issue in the concluding section.

4.3 Nonmarket selection environments

While economists have concentrated their attention on market sectors, research on the selection environment of nonmarket sectors has been undertaken principally by anthropologists, sociologists, and political scientists. This in itself would have led to some significant differences in focus and analysis. But to a considerable extent the differences in analysis appear to reflect real differences in the selection environments.

An essential element in most theorizing about market selection environments is a relatively clear separation of the 'firms' on the one hand, and consumers and regulators on the other. Consumers' evaluation of products — versus each other, and versus their price – is presumed to be the criterion that ought to dictate resource allocation. Firms can be viewed as bidding, and competing, for consumer purchases, and markets can be judged as working well or poorly depending on the extent to which the profitability of a firm hinges on its ability to meet consumer demands as well as or better than its rivals. The viability of an innovation should depend on consumers' evaluation of it.

A hallmark of nonmarket sectors is that the separation of interests between firms and customers is not as sharply defined as in market sectors. The relationship between a public agency, like a school system, and its clientele (students and parents) and sources of finance (mayor, council, and voters), simply does not have an arm's length distance quality that marks the relationship between seller and potential buyer of the new car. Relatedly, the question of how legitimate values are to be determined is much more complex than in market sectors. The public agency is expected to play a key role in the articulation of values, and to internalize these and work in the public interest of its own volition. This is so in many nominally 'private sector' activities, like the provision of medical services by doctors. The doctor is not supposed to make his decisions regarding the use of a new drug on the basis of whether this will profit him, but rather on his expectation of how this will benefit his patients. Further, he is supposed to know more about that than do his patients. This is not to say that in fact interests of firms and consumers always are consonant. In most nonmarket sectors (as in market sectors where competition is lax) the firm has a good deal of discretionary power regarding what it is to provide, and the customer may have little direct power to reward or to punish performance. But in general the appropriate 'control' mechanism over a provider of goods and services in a nonmarket sector is not viewed as competition among providers for the consumer dollar. For these reasons, the motivations of the firms in a nonmarket sector cannot simply be presumed to be monetary profit. This makes analysis of the operative values relating to acceptance or rejection of an innovation difficult. As in the theory of consumer behavior, as contrasted with the text book theory of the firm, tastes matter, these may be hard to analyze, and may not be stable. Even in situations where there is a relatively clear cut goal, and the decision to employ an innovation or not hinges on assessment of efficacy relative to that goal, it has proved hard to identify relevant criteria. Thus, in the Coleman, Katz and Menzel (1957) study of the diffusion among physicians of a new pharmaceutical, the authors did not even attempt to specify quantitatively the ways in which the new product was superior medically to preexisting alternatives. In Warner's study of the decision by doctors to use new chemotherapeutic techniques for the treatment of cancer, in several of the cancer varieties where a significant fraction of patients were so treated, there was no quantitative evidence that the therapy had any effect (Warner, 1974). The physicians made their decisions on hope but on no objective evidence. Friedman (1973), in his study of the acceptance and spread of a certain program of bail reform, was able to identify a few rather specific reasons why the key agencies might find the reform attractive. But the reasons were largely qualitative and it is interesting that, after adopting the reform, there was no real monitoring to check that the programs were performing as hoped. In fact, the performance of the program eroded over time in at least one key dimension, and no one noticed.

Political and regulatory control over firms cannot provide the pervasive, if not always coercive set of value signals and incentives that is provided by consumer sovereignty in market sectors. Thus there is greater room left for autonomous and discretionary behavior on the part of suppliers. However, the employment of regulatory and political mechanisms of governance, as contrasted with consumer sovereignty, means that in many cases several different parties may have to go along before an innovation can be operative. In Friedman's study of bail reform, the police and the courts both had to agree to the proposal, and legislative agreement was necessary where budget was involved. Government agencies often have to gain specific agreement from both political chief executives and legislature before they can proceed with a new program.

Nonseparation of suppliers and demanders leaves little room for firms to compete among each other for consumer dollars. Where there is a single supplying entity – like the United States Postal Office, or the Department of Defense – spread of use of an innovation is a matter of internal decision making constrained and pressured to some degree by the higher order political processes. Where there is a range of suppliers – as in medicine, or in state and local governmental agencies – innovations must spread largely through imitation across the spectrum of noncompeting firms. At the same time there is no incentive for the innovating firm to deter imitation. Organizations that cannot expand into the terrain of others and know that others cannot encroach on their territory have little to gain from preventing others from adopting their successful innovations. Indeed, in most of the sectors under consideration here, there are formal arrangements for cooperation and flow of information across firms. In many, professional organizations set values and judge the merit of new innovations. Professional stamp of approval, and the adoption pattern it stimulates, often are the only criteria for merit available to a non-professional.

Consider the quasi market for the provision of physician services. Without strong constraints afforded by consumers or outside regula-tors, consumer welfare is guarded (perhaps not so securely) largely by professional standards of efficacy of treatment. To assess the efficacy of new treatments doctors consult with each other and apparently aim for professional consensus guided by the judgment of certain key experts. Mohr's study (1969) of the spread of new practices and policies across local public health services reveals a similar professionalism at work. Walker's study (1969) of the lead and lag pattern among state governments in the adoption of new programs indicates the presence of regional groups with intraregional leaders (generally populous, urban, and wealthy states) to which officials in departments in other state governments look for references and models.

Professional judgments are moderated by political constraints on spending limits, and other governmental regulatory processes which impinge on decision making in a more detailed way. Thus in Mohr's study, the speed with which a local public health service adopts new practices was found to be positively related to the extent to which public health professionals were in control of the key office. However, the professional bias toward adoption of new techniques was moderated by-political and budgetary constraints. These, which had to do with the composition and presumably the attitudes of the local 'consuming' populations, did limit, if in a loose way, the innovations that local public health services could afford to adopt. Similarly, Walker's study showed that budgetary constraints imposed by state political systems significantly moderated the proclivity of state officials to adopt progressive programs (read, programs adopted by other states whose judgments they admired).

Crain's study (1966) of the spread of fluoridation across American cities is perhaps the most revealing example of a sector in which the 'firms' have a bias toward adopting an innovation based on notions of professional or technical appropriateness, but consumers tend to resist it. He notes that the spread of fluoridation first occurred quite rapidly, in a context where local health professionals were in charge of the decision. As time elapsed, fluoridation became a more openly political issue, and mayors began to take the decision-making authority out of the hands of the professionals. The spread of fluoridation slowed significantly. Still later, it became common for citizen referendum to become the vehicle for decision. This development brought the spread of fluoridation to a virtual halt.

The pattern in all of these cases is quite different from that in the market sectors studied by economists. It is, however, easy enough to see the same broad elements of modeling that need to be stressed: motivations of the firms in the sector (in general, not characterizable in terms of monetary profit), the ways (if any) in which consumers (often voters) and financers (often legislatures) constrain firm behavior, and the mechanisms of information and value sharing among firms in the imitation process (which is the dominant mechanism by which an innovation spreads).

5. Thoughts on the Effects of Institutional Structure

The preceding two sections sketched some pieces that we hope can be filled in, extended, added to, so that a useful theory of innovation may evolve. The conceptualization has taken very seriously the two critical requirements for a useful theory of innovation that were identified earlier: that innovation be treated as inherently stochastic, and that the formulation be capable of encompassing considerable institutional complexity and variety.

Simply keeping in mind the uncertainty and institutional diversity surrounding innovation can help make thinking about policy issues more sophisticated than has been the norm. Consider, for example, the literature that has evolved on the role of firms with substantial market power in the innovation process.[14] To a distressing extent, that literature has placed the problem in a static frame. Yet Schumpeter, certainly the primal intellectual sources of the current discussion, never viewed the innovation problem statically. Always he had at the center of his analysis that innovation was an uncertain business, that it was important to get new things tried out, to sort out the good from the bad, and that doing this effectively was the principal virtue of capitalistic competition. In his *Theory of Economic Development* (1934), his argument about the inherent monopoly power of successful innovators was concerned explicitly with transient monopoly as a consequence of, and lure for, innovation — not structural monopoly as a base for innovation. There is something more of a structural argument in his *Capitalism, Socialism and Democracy* (1950). But the desirable structure that he proposed is not that of sheltered monopoly. The firms of the *Theory of Economic Development* have grown larger, but 'The Process of Creative Destruction' scarcely is a proposal for a quiet life for monopolists.

Further, most of the analysis has proceeded as if the presence, or lack of it, of large firms with considerable market power were the key institutional difference between economic sectors. In fact, intersectoral differences in the presence of 'large firms' is not a powerful explicand of inter-sectoral differences of productivity growth. If this were the key institutional variation possible, there would be little hope of designing powerful policy. But, of course, there is far more to institutional variation than differences in the average size or market power of firms. In some of the sectors the critical institutions are not firms at all in the ordinary sense (e.g., medical care, garbage collection, etc.). And, even where it can be persuasively demonstrated that large firms can internalize benefits from R&D much more completely than can small firms, an alternative policy would be to establish governmental or not-for-profit organizations to carry on these activities, in lieu of reliance on large firms.

The reason we have stressed uncertainty and institutional complexity and diversity is that these salient attributes of innovation have been ignored in much of the literature concerned with policy towards innovation. This is particularly so in the literature focused on the inter-industry productivity growth puzzle, which we have proposed is perhaps the most important phenomenon that needs to be comprehended if analysis is seriously to contribute to the policy dialogue. However, merely recognizing these aspects cannot carry us very far.

We believe that the two major theoretical proposals we have introduced above – modeling innovation generation as the conditional probabilistic outcome of various R&D strategies, and modeling the fate of an innovation in terms of the workings of a selection environment – have promise of taking us further. For example, the 'selection environment' language seems useful for describing some of the important institutional differences among sectors, and for beginning to explore some of the consequences of these for the inter-industry productivity growth puzzle. Clearly there is considerable variation among sectors in their 'selection' environments, and these differences can affect both the speed and extent of spread of any innovation. It might be conjectured, at first thought, that these differences would influence the level of productivity at any time but not its rate of growth. We think this is wrong for two reasons. First, even if one assumes that the rate of advance of 'best practice' is not influenced by the selection environment, it is not apparent that sectors need be characterized by a constant ratio of average to best practice. In some non-market sectors it is hard to identify strong forces that will prevent that ratio from falling. This suggests that a particularly fruitful arena for investigation is the nature of innovation evaluating and information dissemination systems in non-market sectors (since imitation carries such a large part of the load in these sectors for spreading of the innovation). If we understood these better, it is likely that we would see how to improve upon them.

Second, and even more important, the selection environment feeds back to influence the incentives for R&D by the firms in a sector. We propose that our 'selection environment' concept is much better suited than conventional economic concepts for probing at 'externalities' from R&D. Regarding non-market sectors it is hard to make many generalizations, even tentatively. The most important one is that there is likely to be only a casual connection between the incentives for R&D on the part of public or not-for-profit suppliers, and consumer benefits.

Regarding market sectors, the situation clearly is more complex than economists writing about 'externality' have tended to recognize, and the externalities are inextricably connected with dynamics. For example, consider the comparison we have made elsewhere of selection environments in agriculture and in aircraft production (Nelson and Winter, 1974). The fact that producers of aircraft can expand their plant rapidly, and that imitation is hard, provides great stimuli to the firms to do R&D, perhaps over stimuli. On the other hand, in agriculture, the fact that expansion is slow compared with imitation means that there is little incentive to firms to do R&D. Support for R&D is dependent upon suppliers, and public agencies. The division between 'own' and 'supplier' R&D, and between private and public finance, analyzed in the differential productivity growth literature, clearly reflects underlying structural differences of this sort. The selection environment concept enables these to be analyzed. One can interpret the naive form of the 'Schumpeterian' hypothesis as positing that the returns to R&D are internalized to a greater extent when there are a few large firms than when there are many small ones. But clearly there is a lot more going on than merely differences in size of firms, and further, the sizes of firms are largely endogenous to the system, not a given.

Indeed the explicit dynamic treatment of the selection environment enables one to see potential anti-trust problems, which although they have been discussed verbally, defy formal treatment within traditional models.[15] Where, as in aviation, a significant innovation can give one firm a major advantage over others, where firms can grow rapidly and there are a few limits on firm size, and where imitation is difficult, monopoly structure may develop out of the competitive process itself, and for reasons that textbook discussions focusing on economies of scale may badly obscure. The question of what to do about monopoly structure in the production of civil aircraft should that develop, or what to do about IBM, is extremely complex. But at least the theoretical structure proposed here enables these issues to be seen.

In contrast with discussion of selection environments where the focus was explicitly institutional, there was little explicit institutional discussion in our analysis of R&D strategies. In particular, the discussion of natural trajectories, which we conjectured (interpreting the conjectures of others) were associated with sectors where technological advance has been most rapid, may have given the flavor of 'innate' differences. However, we suggest that it is an open question whether it is inevitable that natural trajectories exist for certain technologies but not for others. We have put forth the proposition that underlying natural trajectories there is a certain body of knowledge that makes the traverse relatively easy, and that in the recent half century formal science has been an important part of that knowledge. The key question then becomes: to what extent are the directions in which science advances inevitable, and to what extent can these be molded by conscious policy.

It is apparent that the evolution of basic scientific understanding has a certain logic, and momentum, of its own. To try to guide that evolution with any precision or to believe that improvements of fundamental understanding can be won simply because the payoffs are high, is foolishness. However, the fields of aerodynamics and applied thermodynamics did not evolve as rapidly and fruitfully as they did merely because they were 'ripe' and groups of academics were interested in them. Rather, they were carefully nurtured and funds and institutions provided for their development. Similarly, both in agriculture and in medicine public institutions and public funds have been established to further the advance of knowledge that feeds into the technologies.

We propose the following. While externalities pervade the innovation process they are greatest in the activities that generate understanding and data. In all of the sectors that have been blessed with strong scientific underpinnings to their technologies, institutions other than the 'firms' in the sector have played a major role in developing that science. In many cases these institutions are 'universities' and the fields defined by academic interests. But in some cases it has been possible to establish institutions that have over the years successfully built up a scientific knowledge relevant to a technology, and which continue to expand that base of knowledge. Study of these cases should be high on the priority list of anyone interested in the problem posed by the great imbalance of technological advance.

But we do not want to press these particular arguments too far. Rather, these kinds of questions, and tentative propositions, are put forth as illustrative of those that arise rather naturally from our proposed theoretical formulation. We hope we have persuaded you that, at the least, there is some promise here for a view of innovation, and a language for talking about innovation, that can integrate a non-trivial share of the present congeries of disjoint research traditions. That would be an important step forward.

Notes

[1] The Joint Economic Committee paper is by Gilpin (1975). King (1974) describes the science-policy thinking at OECD.

[2] Quantitative analyses of the relationship between relative productivity growth, and relative price changes, are contained in Kendrick (1961, 1973) and Salter (1966). Baumol (1967) articulates the malaise about the slow productivity growth industries.

[3] There have been several recent relatively comprehensive reviews of pieces of this literature. Among the best are Mansfield (1972), Pavitt (1971), Freeman (1974) and Kennedy and Thirlwall (1972).

[4] See in Nelson and Winter (1974, 1975, 1976), and Nelson, Winter and Schuette (1976).

[5] The key reference is Hayami and Ruttan (1971). An extension and updating will be published soon by Binswanger and Ruttan (1976).

[6] For the general discussion see Nelson, Peck and Kalachek (1967). Baumol (1967) has stated the hypothesis about the service sectors. See Rosenberg (1972), David (1974), Hughes (1971) and Levin (1974) regarding mechanization and economies of scale.

[7] For good recent reviews and criticism see Binswanger (1974) and Nordhaus (1973).

8   Note that the concept of a strategy employed here is *not* the same as the strategy concept used in statistical decision theory or game theory. In general a strategy, as we are employing the term, will not involve an explicit ex-ante solution to what will be done under all possible contingencies. Nor is our kind of strategy derived from any explicit maximization calculations.

9   For a rich and sophisticated discussion see Simon and Newell (1972).

10   For a discussion see Pavitt (1971) and Freeman (1974).

11   See Miller and Sawers (1968) and Phillips (1973).

12   An earlier version of this section was published in Nelson and Winter (1975) and development process.

13   For a review see Mansfield (1973).

14   For a fine review of that literature sec Kamien and Schwartz (1975).

15   For a discussion see Phillips (1973).

References

Baumol W. (1967), 'The Macro Economics of Unbalanced Growth,' in *American Economic Review*, 57(June): 415.

Binswanger H. (1974), 'A Micro Economic Approach to Induced Innovation,' in *Economic Journal*, 84 (December): 940.

Binswanger H. and Ruttan V. (1976), *The Theory of Induced Innovation and Agricultural Development*, Baltimore: John Hopkins.

Brown M. and Conrad A. (1967), 'The Influence of Research and Education on CES Production Relations,' in Brown M. (ed.), *The Theory and Empirical Analysis of Production*, Columbia University Press for NBER.

Coleman J., Katz E., and Menzel H. (1957), 'The Diffusion of an Innovation Among Physicians, in *Sociometry*, 20 (December): 253.

Crain R. (1966), 'Fluoridation: The Diffusion of an Innovation Among American Cities,' in *Social Forces*, (June): 467.

David P. (1974), *Technical Change, Innovation and Economic Growth*, Cambridge: Cambridge University Press.

Evanson R. and Kislev Y. (1975), *Agricultural Research and Productivity*, Yale University Press.

Freeman C. (1974), *The Economics of Industrial* Innovation, Harmondsworth: Penguin.

Friedman L. (1973), Innovation and Diffusion in Nonmarkets: Case Studies in Criminal Justice, Ph.D. Dissertation, Yale University.

Gilpin R. (1975), Technology, Economic Growth and International Competitiveness: A Report Prepared for the Joint Economic Committee of the U.S. Congress, July 19, Washington: USGPO.

Grabowski H. and Mueller D. (1972), 'Managerial and Stockholder Welfare Models of Firms Expenditures,' in *Review of Economics and Statistics*, 54 (February):9.

Griliches Z. (1973), 'Research Expenditures and Growth Accounting,' in Williams B. R. (ed.), *Science and Technology in Economic Growth*, New York: Halsted Press.

Habakkuk H. J. (1962), *American and British Technology in the 19th Century*, Cambridge: Cambridge University Press.

Hayami Y. and Ruttan D. (1971), *Agricultural Development: An International Perspective*, , Baltimore: John Hopkins.

Hughes W. (1971), 'Scale Economies and Electric Power,' in Capron W. (ed.), *Technical Change in Regulated Industries*, Washington: Brookings.

Jcwkes J., Sawers D. and Stillerman R. (1961), *The Sources of Innovation*, New York: Norton.

Kamien M. and Schwartz N. (1975), 'Market Structure and Innovation: A Survey,' in *Journal of Economic Literature*, 13(March):1.

Kendrick J. (1961), *Productivity Trends in the United States*, Princeton University Press for the NBER.

Kendrick J. (1973), Postwar Productivity Trends in the United States, 1948-1969,Princeton U. Press for the NBER.

Kennedy C. and Thirlwall R. (1972), 'Surveys in Applied Economics: Technical Progress: A Survey,' in *Economic Journal*, 82(March): 11.

King A. (1974), *Science and Policy: The International* Stimulus, Oxford University Press.

Leonard J. W. (1971), 'Research and Development in Industrial Growth,' in *Journal for Political Economy*, 79 (March and April): 232.

Levin R. (1974), Technical Change, Economies of Scale, and Market Structure, Ph. D Dissertation, Yale University.

Mansfield E. (1972), 'The Contribution of R & D to Economic Growth in the U.S.,' in Science, 175(February): 477.

Mansfield E. (1968), Industrial Research and Technological Innovation, New York: Norton.

Mansfield E. (1973), 'Determinants of the Speed of Application of New Technology,' in Williams B. R. (ed.), *Science and Technology in Economic Growth*, New York: Halsted Press.

Meyer J. and Kuh E. (1957), *The Investment Decision: An Empirical Study*, Harvard University Press.

Miller R. and Sawers D. (1968), *The Technical Development of Modern Aviation*, Routledge and Keegan Paul.

Mohr L. (1969), 'The Determinants of Innovation in Organization,' in *The American Political Science Review* , 63: 111.

Mueller D. (1967), 'The Firms Decision Process: An Econometric Investigation,' in *Quarterly Journal of Economics*, LXXXI(February): 58.

Nasbeth L. and Ray G. (1974), *The Diffusion of New Industrial* Processes, Cambridge University Press.

Nelson R. and Winter S. (1974), 'Neoclassical Vs. Evolutionary Theories of Economic Growth,' in *Economic Journal*, 84 (December): 886.

Nelson R. and Winter S. (1975), 'Factor Price Changes, and Factor Substitution in an Evolutionary Model,' in *Bell Journal of Economics*, 6 (Autumn): 466.

Nelson R. and Winter S. (1975), 'Growth Theory from an Evolutionary Perspective: The Differential Productivity Growth Puzzle,' in *American Economic Review*, 65(May):338.

Nelson R. and Winter S. (1976), 'Dynamic Competition and Technical Progress,' in Balassa B. and Nelson R. (eds.), *Private Incentives, Social Values and Public Policy:Essays in Honor of Mlliam Fettner*, Amsterdam: North-Holland.

Nelson R., Peck M. J., and Kalachek E. D. (1967), *Technology, Economic Growth and Public Policy*, Washington: Brookings.

Nelson R., Peck M. J., and Kalachek E. D. (1967), *Technology, Economic Growth and Public* Policy, Washington: Brookings.

Nelson R., Winter S., and Schuette H. (1976), 'Technical Change in an Evolutionary Model,' in *Quarterly Journal of Economics*, 90 (February): 90.

Nordhaus W. (1973), 'Some Skeptical Faults on the Theory of Induced Innovation,' in *Quarterly Journal of Economics*, 87(May): 208.

Nyman S. and Silberston A. (1971), 'An Approach to the Study of the Growth of Firms,' Paper presented at Second Conference of Economics of Industrial Structure, Holland.

Pavitt K. (1971), Conditions of Success in Technological Innovation, Paris: OECD.

Phillips A. (1973), *Technology and Market Structure,* Heath Lexington.

Piori M. (1968), 'The Impact of the Labor Market upon the Design and Selection of Productive Techniques within the Manufacturing Plant,' in *Quarterly Journal of Economics*, 82 (November): 602.

Raines F. (no date), 'The Impact of Applied Research and Development on Productivity', *Washington University Working paper*, No. 6814.

Rosenberg N. (1969), 'The Direction of Technological Change: Inducement Mechanisms and Focusing Devices,' in *Economic Development and Cultural Change*, 18(October):6.

Rosenberg N. (1972), *Technology and American Economic Growth*, New York: Harper Torch Books.

Salter W. E. H. (1966), *Productivity and Technical Change*, Cambridge University.

Schmookler J. (1952), 'The Changing Deficiency of the American Economy, 1869-1938,' in *Review of Economics and Statistics*, 34(August): 214.

Schmookler J. (1966), *Innovation and Economic Growth*, Harvard University Press.

Schumpeter J. (1934), *The Theory of Economic* Development, Harvard University Press.

Schumpeter J. (1950), *Capitalism, Socialism and* Democracy, New York: Harper and Row.

Science Policy Research Unit (SPRU) (1972), *Success and Failure in Industrial Innovation*, Center For Study of Industrial Innovation, London.

Setzer F. (1974), Technical Change over the Life of a Product: Changes in Skilled Inputs and Production Processes, Ph.D. Dissertation, Yale University.

Simon H. and Newell A.(1972), *Human Problem* Solving, New York: Prentice Hall.

Terleckyj N. (1974), *The Effect of R & D on Productivity Growth in Industries*, National Planning Association, November.

Walker J. (1969), 'The Diffusion of Innovation Among American States,' in *American Political Science Review*, 63 (September): 880.

Warner K. (1974), *Diffusion of Leukemia Chemotherapy*, Ph.D. Dissertation, Yale University.

Acknowledgement

This paper was originally published in *Research Policy*, Vol. 6 (1977): 36-76. We thank the authors, the editors of *Research Policy* – Ben Martin and Martin Bell, and the Elsevier Publishers for giving the right to reprint it in AJSTID - Editors.

This paper was supported partly under NSF Grant No. GS 35659 and partly under a grant to Yale from the Sloan Foundation. An earlier version of part of this paper was presented at a conference in Cambridge, Summer 1975.

African Journal of Science, Technology, Innovation and Development
Vol. 1, No. 1, 2009
pp. 53-75

Building Systems for Innovation "Take off" in African Economies

Abdelkader Djeflat*

Abstract

While catch up theory assumes that innovation takes place in a linear model and that path dependant trajectory characterize the innovation process, "innovation take off" rests on the premise that innovation systems need strong policy impulses from government for innovation to effectively take place. Innovation take off is the prerequisite for innovation systems to operate in a conventional manner. In many developing countries, innovation systems construction takes place in a very specific environment characterized by the rise of a small and medium scale enterprises (SMEs) sector but with little experience in the fields of R&D, relatively weak industrial performances in terms of productivity, and high level of obsolescence both in terms of human resources and equipment. While the approach in terms of innovation systems is real and attracts a great deal of attention from policy makers, very little work has been done on the way African countries could engage in a real take off policy of their innovation system in a sustainable manner. This contribution addresses the fundamental question of how innovation takes off in late industrializing countries such as African countries, both in terms of policies and conceptual framework. It draws from the Algerian experience and takes further the conceptual and theoretical approach that have been developed in earlier works on industrial technical centres as a way to trigger off innovation.

Keywords: Innovation, R&D, innovation take off, catch up, industrial technical centres, less developed countries, Africa.
JEL Classification: O31, O32, O34

---

*   The Maghtech Network, Lab. CLERSE/CNRS, University of Lille1 - France; Email: Abdelkader.Djeflat@univ-lille1.fr

Abdelkader Djeflat/AJSTID Vol. 1, No. 1 (2009), pp. 53-75

1. Introduction

While catch up theory assumes implicitly that innovation takes place in a linear path and that markets play a key role, "innovation take off" rests on the premise that innovation systems need strong policy impulses from government for innovation to effectively take place. Innovation suffers from a real crisis, not a maturity crisis known in certain advanced crisis but a crisis of birth in the framework of an innovation life cycle to be determined. If National Systems of Innovation (NSI) in the developed world are considered to be in an advanced stage, those of the developing world are rather in a primitive stage (Gu, 1999).

Innovation take off is the prerequisite for innovation systems to operate in a conventional manner. Innovation systems theory has long been characterized by the difficulty in using it in system construction (Lundvall *et al.*, 2002), yet many less developed countries (LDC) seem to have resorted to this approach to build their innovation systems. Good examples are found among the BRICS countries (Brazil, Russia, India, China and South Africa).

In many other LDCs, like the Maghreb countries (Algeria, Tunisia, Morocco), innovation systems construction takes place in a very specific environment characterized by privatization of public concerns, the rise of a strong small and medium enterprises (SMEs) sector, but with very little experience in the fields of R&D and innovation, and a relatively weak industrial sector in terms of performances, suffering notably from high levels of obsolescence both in terms of human resources and equipment.

While the approach in terms of innovation systems is real and attracts a great deal of attention from policy makers in these countries, many of them are seeking the best and quickest way to start off innovation notably in their industrial sector, looking at the advanced world experience and trajectories but more and more at the BRICS experience. Yet when looking at these countries, where innovation is effectively taking place, a variety of situations and trajectories exist.

This paper addresses the fundamental issue of innovation "take off" in late industrializing countries such as the Maghreb countries, both in terms of policies and conceptual framework. The issues raised relate to innovation dynamics in the early stage of system construction and notably: What are the characteristics of innovation in the pre take off stage? How do African countries compare with emerging countries and the advanced world? How does take off relate to catch up and to what extent is it more relevant to African countries than catch up? Innovation systems at the takeoff have special characteristics and notably have their own trajectories: what are they and do they help the construction of innovation systems for takeoff? This raises two fundamental issues: firstly, what are the modes of construction of innovation systems in the take off stage and what steps can be figured out, and secondly, what is the driving engine for innovation take off?

The paper is an outcome from a major work conducted in 2007/2008 for the European Union (MEDA II) to support innovation systems building in Maghreb countries. It consists of five sections: section 2 discusses the concepts of innovation take off, and how it relates to the catch up approach; section 3 concentrates on the modes of construction of innovation systems in the takeoff stage; section 4 examines industrial technical centres as possible engines for innovation take off; and finally section 5 draws some conclusions.

2. Innovation Dynamics at the Take off Stage

2.1 The necessary take off stage in innovation dynamics

It is a fact easily documented that major efforts have been made by a number of developing countries to launch innovation, pressurized as it were, by innovation based competition. It is also a fact that innovation output remains relatively poor in these countries. All kinds of indicators such as R&D expenses as percentage of Gross Domestic Product contribute to show that. Some factors, listed below, can be used to explain this situation.

Technological backwardness of Africa compared to other countries: As noted elsewhere, many countries, including the developing ones, remain low R&D performers and continue to lag behind on other aspects of great relevance for creation and exploitation of knowledge in the contemporary world (Fagerberg *et al.*, 1999; Fagerberg and Godinho, 2004). The gap is clearly shown when looking at conventional indicators. The number of scientists in R&D in sub-Saharan Africa in 2001 reached 3,193 individuals, which amounts to 0.3% of the total number of researchers in the developing world and 0.1% of the total number of researchers in advanced countries (see Table 1).

Table 1: State of R&D in the world (2001)

Countries and Regions	Number of Engineers involved in R&D	Total R&D (%GNP)	Performance (%): Production sector	Performance (%): Higher education
Developed countries	2 704 205	1.94	53.7	22.9
Developing Countries	1 034 333	0.39	13.7	22.2
Sub-Saharan Africa	3 193	0.28	00	38.7
North Africa	29 675	0.40	NA	NA
Latin America	107 508	0.45	18.2	23.4
Asia (excluding Japan)	893 957	0.72	32.1	25.8
World	4 684 700	0.92	36.6	24.7

Source: Lall and Pietrobelli (2003); Note: NA: Not Available

Moreover, R&D funded by companies in East Asia is 250 times more than in African countries (not including South Africa), 25 times more than in Latin America and twice more than in transition economies. There are very limited R&D expenditures in sub-Saharan Africa and the situation does not seem to have improved significantly as a whole at the beginning of the years two thousand (Economic Commission for Africa, 2004).

Limited absorption capacity

Major efforts were made, however by some countries in recent years to boost R&D and expenditures increased in a relatively short time from 0.3% to 0.7%, and even to 1% of GDP, notably in North Africa (Djeflat et al., 2007). Often this effort associated with some kind of policy has not produced any significant result showing that, financial thrust, by itself does not get innovation off the ground within a reasonable time. On the contrary, too much funding has created some undesirable side effects. Thus one of the important problems is the "absorptive capacity" of these countries.

Table 2: Planned and allocated budget in the five year period from 1999 to 2003 in Algeria (in Dinars - billion)

Budget	1999	2000	2001	2002	2003	Total
Planned budget	21.15	31.21	33.66	36.38	36.38	158.78
Effectively used budget	5.1	4.1	5.1	4.6	5.6	24.5
Used/planned (%)	24.11	13.13	15.15	12.64	15.39	15.43

Sources: National Syndicate of Researchers (2006), cited by H. Khel-faoui *Les technologies de l'information , politique de recherche et d'innovation et politique scientifique,* Report to the Ministry of Urban Planning and Environment, Algeria, p. 12 ; *Note:* 1US$ = 74 Dinars.

As shown in Table 2, less than 16% of the allocated funds in Algeria could be absorbed by the national research system. Many factors could explain this (Djeflat et al., 2007). The first, relates to the weakness of the human element, which remains a key element, coupled with a poor institutional and incentive regime. FTE (Full time equivalent) researchers mobilised represents less than 10% of what could be mobilised. Thus all ingredients appear to exist without the expected performances, unlike in advanced countries where similar effort would have yielded satisfactory results. Critical mass, defined as the minimum level required of a mix of human, financial and institutional ingredients, appears to be one of the key issues. The importance of absorptive capacity has been highlighted in several contributions as a complementary necessity to knowledge creation in relation to technologies acquired abroad (Mowery and Oxley, 1997; Kim, 1997), as a prerequisite to the learning process at the firm level, which necessitates notable intangible investments (Cohen and Levinthal, 1989), and as capacity to create new knowledge and to search and select the most appropriate technology (Narula, 2004). This lack of absorptive capacity on the part of firms could also be the result of the lack of applicability of technological knowledge to local conditions, making collaboration with underdeveloped socio-economic structures virtually impossible in a non-system perspective (Szogs, 2004). The difficulties met reflect the crisis which the national absorption capacity shows. This can be identified with several factors (Narula, 2004): weak or missing basic infrastructure, (means of communication, electricity, health, and basic education), advanced infrastructure (universities, research institutes, firms-domestic and foreign affiliates) and formal and informal institu-tions (intellectual property rights regime, taxation, incentive system and partnerships). It is also a characteristic of the difficult take off which raises slightly different issues than the catch up problem.

The problem in sub-Saharan Africa can be better highlighted by comparing the countries in the region to a newly industrializing country such as South Korea. Figures show a relatively high absorptive capacity in terms of R&D expenditure, in this case, leading to a successful take off of its innovation system. In terms of industrial research, South Korea is in the sixth place among OECD countries,[1] with R&D expenditure reaching US$14.43b per year. Globally, industrial R&D reached 2.64% of GDP in 2003 (see Table 3). In more details, enterprises account for 70% of the total R&D expenditure: half of these come from the electronics sector, followed by the automotive (14.8%) and chemical industries (10%).

Table 3: Industrial R&D expenditure as a percentage of GDP in South Korea (1970-2003)

Years	Percentages
1970	0.39
1975	0.42
1980	0.56
1985	1.52
1990	1.87
1994	2.44
1995	2.37
1996	2.42
1997	2.48
1998	2.34
1999	2.25
2000	2.39
2001	2.59
2002	2.53
2003	2.64

Source: Ministry of Science and Technology & Korea Institute of Science and Technology Evaluation and Planning (2004) cited by Jay-Ik Choi, Vice President of the Korean Industrial and technological Association: les activités de R&D dans les entreprises Coréennes , ANRT Répères sur l'innovatin en Corée, Paris.

Weak irregular and uncertain R&D result. As noted elsewhere, several African countries have relatively weak R&D results and are weak on the creation and exploitation of knowledge in the contemporary world (Fagerberg *et al.*, 1999; Fagerberg and Godinho, 2004). Figures from the Algerian Patent Registration Office show the weak performances of R&D and the relatively slow and difficult progress that has been made. These are signs of the difficult take off of innovation performances as shown by the number of registered patents by residents. The peak reached in 1965 did not exceed 60 patents. During the ten years between 1983 and1993, there were less than 10 patents on average registered each year (Figure 1).

Figure 1: Evolution of the number of registered patents by residents in Algeria

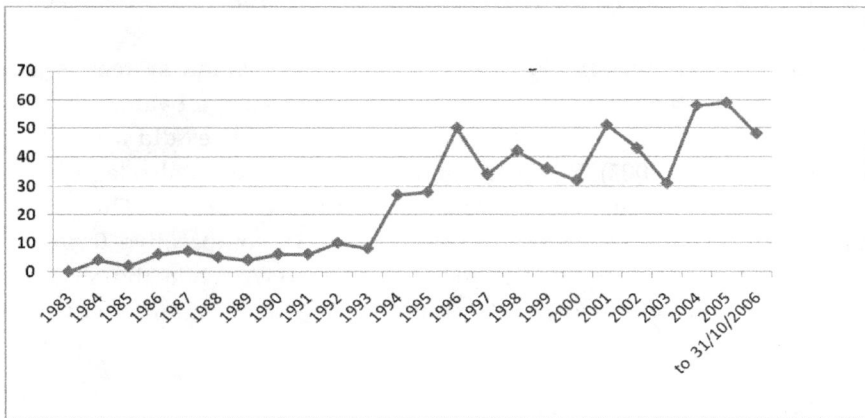

Source: Djeflat *et al.*, (2007)

Comparatively in South Korea, the total number of registered patents has been growing at a relatively high pace, having grown by 100 times between 1981 and 2000 (see Table 4). The number of registered patents in the United States Patent Office (USPTO) jumped from 943 patents in 1994 to 3944 in 2003.

Table 4: Number of registered patents in South Korea (1981-2000)

Patent Type	1981	1985	1990	1995	2000
Total Number of registered patents (T)	1808	2268	7762	12512	34956
Patents registered by Koreans (K)	231	349	2554	6575	22943
K/T (%)	12.8	15.4	32.9	52.5	65.6

Source: Korean Intellectual Property Office, cited by Yim D. S. (2004), *Korea's National Innovation System and the Science and Technology Policy,* Science and Technology Policy Institute (STEPI), p.10.

Furthermore, the difficult take off of the innovation system is corroborated by the high proportion of individual innovators, which represent 84% of the total, while the share of enterprises did not exceed 9% (Djeflat *et al.*, 2007). Research centres and universities also lag behind with only 6%, showing a real crisis of institutional research. This is also the syndrome of stagnation, i.e. pre take off. In comparative terms, in France the proportion of institutions reached 68% while the share of individuals did not exceed 16%, another indicator of an innovation system in the stage of maturity.

2.2 From take off to catch up

The take off stage stems much from the take off theory of Rostow (1960). Various critics have long pointed out the main weakness of this model, namely its linearity and we would go along with this, knowing that standard theory has largely failed to convey the complex and interactive nature of the innovative activities. One can argue in this particular case that take off is not at the beginning of the process but at the end of a complex process, whereby preliminary conditions incorporate a set of physical, financial, human, and institutional as prerequisites for innovation to take place, but also, from an evolutionary perspective, complex, cumulative and path dependant interactions. These preliminary conditions need further clarifications. The experience of industrialized countries show that they benefit from rich learning interactive spaces in the sense that several productive encounters take place between those that need knowledge and those that could interact with them. Countries in the South can have only poor learning interactive spaces as a result of the scarcity of these interactions (Arocena and Sutz, 2003). NSI functions in the wake of diversity and variety (Johnson, 1992), of uncertainty resulting from their evolution according to a selected trajectory, selectivity and path-dependency or historicity and finally of irreversibility (Niosi *et al.*, 1992). It is interesting to raise the question at this stage of the relevance of catch up theory for these countries.

Catch up, while being appropriate for several emerging countries seems of little relevance in a context where innovation is largely inaccessible and where the innovation take off is still problematic. Similarly, catch up appears more appropriate to the process of catching up world leaders (Johnson and Lundvall, 2003), which is remote from the preoccupations of the majority of African countries with the exception of South Africa where performing industry exists (Avnimelech and Teubal, 2006) . Catch up theory presupposes implicitly the existence of a structured, complete and fully operational NSI. Reality shows that we could not assume the existence ex-ante of NSI and the limited work done on Africa give little evidence of its existence. Often NSIs exist in a preliminary form, are unstructured, disorganized and fragmented. For example, studies on North and West Africa show the existence of uncoordinated components largely disconnected from public policies (Djeflat 2003, Casadella 2006, Carré 2002). Similarly, sectorial innovation systems cannot be assumed ex ante as shown in studies in the agro-food industries in North African countries (Ait Habouche *et al.*, 2004) knowing that this is a prerequisite for catch up to take place (Avnimelech and Teubal, 2006).

Catch up implies the existence of effective demand for R&D and innovation services while reality shows that this demand remains still very weak in spite of the opening of these economies and the pressure of competition. Development of a firm, a regional or a national innovation system could not possibly evolve unless a substantial demand exists for new products and services and subsequently for R&D activities (Nielsen, 2005).

Take off appears to be more appropriate for several reasons:

Firstly, it is more adapted to the current state of economies where innovation systems are still in the construction stage in which most African countries are (Muchie *et al.*, 2003), characterized mostly by incomplete NSI, weak or missing links and weak incentive systems (Narula, 2004, Djeflat, 2004). They suffer from a deficit of interactions between the main components (Casadella, 2006) and pronounced rent-seeking on the part of the main actors (Djeflat, 2004) to the extent that they can be considered as nonexistent (Arocena and Sutz, 2003).

Secondly, it put the emphasis on the necessity for certain preliminary conditions to reach a critical mass required for the take off to take place. Consequently, take off is more appropriate in the sense that system construction requires that certain conditions are fulfilled. Among these, the creation of learning capacities appears to play a key role in a bottom up strategy (Casadella, 2006).

Thirdly, take off appears to be a vital stage in the emerging countries that have evolved in the superior stage, which is catch up stage. This can be seen in Brazil, China, South Korea and India. Finally, take off stage needs a strong state support particularly regarding the institutional dynamics in terms of regulations, salaries and incentive systems, public procurements, and so on, while the conditions of market dynamics are being put in place in transitional economies such as Maghreb countries for example. This is the reason we talk about system construction or promotion (Lundvall *et al.*, 2002) as a substitute to its *reproduction*, which is more appropriate for emerging countries.

2.3 Characteristics of Innovation systems in the stage of take off

Unlike NSI of the North, innovation systems in the South cannot be assumed to have similar characteristics (Edquist, 1997). The diffusion of the NSI concept in the South is possible only if two requirements mentioned previously are fulfilled and well understood: the character ex-post of NSI and the construction of technological capacity. This gives NSIs in the South their own specificities and the heterogeneity of their trajectories, and explains to some extent the various stages of development they have reached. While innovations in the *high-tech* sectors are relatively sophisticated and based on science and radical innovation, NSIs in the South reflect routine perspectives of techniques, where learning by using predominates, R&D activities are not clearly defined and formally articulated within enterprise strategies (Arocena and Sutz, 1999; 2002). Problems of vulnerable and unstable macroeconomic environment are also quite important (Cassiolato and Lastres, 1999) and constitute real obstacles for innovation to the extent that we could talk about *national system of inertia* (Hobday 1995; Hobday *et al.*, 2004). Networks between industry and R&D tend to be absent (Wangwe, 2003). The inability to put in place mechanisms of collective learning is a marked feature, knowing that without collective learning, it is difficult to talk about the existence of innovation system (Archibugi et al. 1998). In this respect, the innovation gap as defined by Arocena and Sutz (2000), goes beyond access to technological know-how and entices three fundamental problems (Oyelaran-Oyeyinka, 2004). The first one is the inability of local institutions to interact with productive entities in order to guarantee a sufficiently autonomous technological dynamics. The second one relates to the difficulty in the building of local knowledge through the tacit knowledge of small African structures which are unable to face new needs in an unstable competitive environment. The third one relates to the repetitive techniques of learning through imitation, leaving relatively little room for opportunities to renew and modernize enterprises' knowledge. Local knowledge is used by institutional actors in charge of the promotion of real endogenous capabilities. Finally NSI in the South are relational, normative and built ex post (Arocena and Sutz, 1999; 2002),

While trajectories constitute a key concept in the catch up theory (Malerba, 2004), in take off approach, they take a particular meaning. Trajectories in African countries do not seem to be continuous and linear. They are often broken trajectories of sectors which accumulate know how and knowledge, which manage the process of gathering the necessary conditions for takeoff but manage to regress, and de-accumulate through de-learning. The sources of de-accumulation are numerous: instability of competencies, and relatively high and continuous labour and employee *turnover* (Oyelaran-Oyeyinka, 2004), the attractiveness of more lucrative sectors often in the tertiary sector, the exodus of competencies and the effects of structural adjustment programmes. The latter have been major factors of de-accumulation as a result of dismantling numerous public enterprises in the manufacturing sector and the laying off of their employees (Djeflat and Boidin, 2002; Djeflat, 2004; Casadella, 2006).

The declining investments in industry as a result of falling external revenues has contributed a great deal to these broken trajectories. This interrupted learning process helps to explain the weaknesses of learning and managerial capacities in the strict sense of the word (Lall and Pietrobelli, 2002; Johnson and Lundvall, 2003). Liberalism also derives from making less effort in R &D to benefit from ready-made technology due to more open economy and additional facilities for importing (Naclerio, 2004). This is not to be assimilated to what Johnson (1992) calls de-learning, the capacity of forgetting (Mytelka, 2000) so indispensable to technical change at the firm level. Competition for local resources coming from basic and more urgent needs in terms of health, food and basic infrastructures contribute also to adjourning or marginalizing research and innovation programmes and policies. Finally, 'rent-seeking' which characterizes many African economies tends to strengthen existing structures and practices, leaving very little room for creativity, invention or innovation.

Consequently, two major conclusions can be drawn: the first one, is the need to construct an innovation system more appropriate to the take off stage whose main characteristic is to exercise a relatively strong push for an effective demand for R&D products and services to emerge; and the second one is the necessity to build an innovation system which could move the whole apparatus from take off to catch up at a latter period.

3. Modes of Construction of Innovation Systems in Take Off Stage

The absence of structured and functional NSI in LDCs has been high-lighted in several contributions partly as a result of building them ex post (Lundvall *et al.*,2002; Casadella, 2006; Carré, 2002; Djeflat, 2003). Similarly, several attempts are made to characterize a more appropriate way to build innovation systems which can be considered sometimes as a substitute and at other times as an intermediary stage prior to the building of a full-fledged NSI. Of these, we can identify from the literature notably National System of Science and Technology (Masinda, 1998), National System of Technological Learning (Viotti, 2002, 2003), National System of Technological Capacities (Lall, 2000), National Capacity of Absorption (Narula, 2004), National Technological System (Lall and Pietrobelli, 2003), National System of Innovation and Construc-tion of Competencies (Lundvall *et al.*, 2002; Muchie *et al.* 2003), National System for Development (Edquist, 2001), National System of Construc-tion of Competencies (Casadella, 2006) and finally Technical Support System For Technological Innovation (Djeflat, 2006).

3.1 Steps in the construction of NSI in the take off stage

While a systematic approach in the construction of NSI could not be identified in the take off stage in the literature, there are components in some approaches, which could help figure out stages in this construction process. Three approaches can be briefly mentioned:

In the first one, a descriptive model links construction process to stages of development (Liu and White, 2001). Six activities related to the innovation process are identified: research, production, interactions, coordination, control mechanisms and education. Their analysis rests on the distribution of activities in the innovation dynamics. Liu and White (2001) apply this method to analyze the Chinese NSI throughout the various stages of development and highlight differences in the institutional framework, in organization and in performances of NSI of Communist China and Democratic China.

A second approach which goes along these lines has recently been put forward (Casadella, 2006) which identified four stages:

Stage 1: non action in the pre take off stage, a non-system perspective (Szogs, 2004).

Stage 2: decentralized improvement of knowledge - *National System of Construction of Competences* based on the mobilization of local knowledge, and use of knowledge in the perspective of improving learning capacities and guaranteeing of learning effort (Casadella, 2006).

Stage 3: Centralized improvement of knowledge through NSI in the broad sense, where Technological Learning (Viotti, 2002) involving diffusion of foreign technology, preliminary R&D and other efforts of technological learning take place.

Stage 4: Knowledge creation through NSI in the strict sense of the term accompanied by the creation of new knowledge and full-fledged R&D activities.

A third approach takes into account the various intermediary stages:

Stage 1: pre take off stage characterized by a *national system of inertia* (Hobday, 1995; Hobday *et al.*, 2004).

Stage 2: creation of learning capacities through bottom up strategies (Casadella, 2006) and adequate learning interactive spaces (Arocena and Sutz, 2003).

Stage 3: creation of national capacity of absorption (Narula, 2004),

Stage 4: completion of the National System of Construction of Competencies (Casadella, 2006)

Stage 5: Technical Support System for Technological Innovation (Djeflat, 2006)

Stage 6: finally a full fledged innovation system.

These approaches are tentative and they need further exploration and greater analytical depth from a theoretical point of view. They also need solid empirical investigations that are specific to African countries, taking into consideration their specificities but also the variety of situations we come across in the continent.

3.2 The driving engine for take off

While the various stages for take off are important for understanding the mechanisms, the driving force behind take off is of an utmost importance. Indeed take off would not occur unless an important driving force is in place as in the case of emergent countries. There are several elements which could play this role. Of these, three major elements can be indentified: the guarantee to innovation effort, public procurement and industrial technical centres.

The guarantee to innovation effort: the development of learning capacity and the acquisition of technological capacity are closely related to the opportunities to apply these capacities to solve technical and technological problems (Casadella, 2006). These opportunities are also fundamental to develop innovation capacities (Arocena and Sutz, 1999). Indeed if opportunities to build an appropriate, tailor made and context specific solutions to local problems, are not given, it would be difficult to get innovation off the ground. Several examples exist in the industrial experience of African countries and particularly those that have access to surplus income to resort to foreign companies, design offices, laboratories and centres of research for any kind of problems faced, missing thus valuable opportunities for local innovation capabilities. In other words, innovation take off cannot be triggered off through conventional market mechanisms, but needs a relatively strong institutional support made of a robust legal system, a good incentive system, and political engagement on the part of actors of the innovation set up (Lundvall, 1992;Von Hippel, 1988). Thus a guarantee to innovation effort is necessary. This effort is made by individuals and institutions at various levels: micro economic, meso economic and macroeconomic.

Public procurement: Public procurement can play a vital role in the face of market failure in the take off of innovation system. It is acquiring more and more importance both as a policy instrument and a concept to be further investigated (Edquist et *al.*, 2000). The actions of government were critical in the emergence of innovative technologies and services in most advanced countries. Public demand for knowledge and innovation exist even in the prospect of privatization. Innovation projects need a broad range of products and services, and an important capital lay out. The role of the State is paramount. This role is even more important in African countries where the technological base of the private sector is relatively weak. The State plays an important push role before the private sector and market become strong enough to take over. A public- private partnership to build this sustainable demand for innovative products is also a key to this process.

4. Industrial Technical Centres as Engine for Take Off

Industrial Technical Centres (ITC) appear to play a significant role in innovation take off particularly for SMEs as shown in recent contributions (Devalan, 2006; Djeflat, 2008). Their main aim is to cater for specific technological demands coming from industry which is usually heterogeneous - composed of SMEs and big firms. SMEs usually have relatively little means to allocate funds to proper R&D and also to testing and control, which require relatively high investments and competencies. In industry where the majority are big concerns, there may be less need for ITCs, knowing that they usually are well endowed with funds, human competences and accumulated experience, proper R&D facilities and proven innovative activities. Basically, ITCs constitute valuable entities for technology transfers which have both human capabilities and material means in terms of finance and equipment to cater for specific needs of a company, an industry or a sector.

The success of ITCs rests mainly on their link position, which is relatively unique and the notion that "the industry network is more than the sum of the capabilities of firms" (Foss, 1999). ITCs are the result of a partnership between the public and the private sectors at the crossroad of research and industry. They are also at the crossroad of several disciplines as 'technology assemblers' capable of analysing, adapting and transferring useful technologies for the design of new products or methods of production. Finally, ITCs constitute a national network located in between regions.[2] It has been frequently underlined that those institutions can go well beyond firms or industrial agglomerations (Lundvall, 1992; Kirat and Lung, 1999; Maskell, 2001). The major difference between ITCs and consulting or industrial services bodies is that the latter are capable of undertaking studies for the profession as a whole to upgrade technological capabilities and the way they are financed by members of the profession.

It is also suggested that the potential of ITCs for learning can be linked to their role as providers of services partly duplicating those provided by or in some firms. According to Mota and De Castro (2003) –

Some similarity of capabilities in specific areas can help to maintain a great proximity to and relevance for the context in which firms operate, and thus facilitate the processes of dissemination of knowledge in the industry. It also suggests that the motivations and benefits perceived by firms and, in general, the relevance of sharing experiences in this context should be seen in the context of firm's specific and idiosyncratic trajectories. The involvement of firms in technological centre activities may be a means for them to directly or indirectly access the experiences of other firms and individuals in an industrial system. The potential for generation of benefits may be associated to the perceived or real similarity between the activities that may be accessed through those centres and those carried out within the firms themselves, especially the activities that are perceived as most relevant in the context of their relationships with their clients and or suppliers.

Three central roles of institutes are highlighted: they are a learning partner for industry, they help to increase absorptive capacity, and they constitute a flexible repository in the innovation system by helping firms in peak periods and by reducing the pressure on universities through assisting in teaching and supervision.

Their inscription at the local and territorial level gives them extra closeness to university and higher education system as a whole and thus they are part of the National Innovation System.

Some of the services rendered may not be classified as R&D activities, but at the take off stage, they can be considered as important by-products and complementary services to R&D and in this respect contribute to broaden the span of NSI activities at this stage to include non traditional and non conventional and yet vital activities for innovation to effectively take place.

As shown in other contributions, ITCs can be considered to possess "unique" competencies and hold high quality which drives firms that lack skills, capacity in R&D and equipment to conduct in house R&D projects to use them (Nerdrum and Gulbrandsen, 2007): ITCs constitute, in this respect an important contribution to firms in their search for innovation. Their contribution rests on their major tasks: they represent a *learning partner* to enterprises. Learning happens primarily through collaborative projects, collective R&D, public programmes oriented towards general interest. Secondly, they contribute to increasing *absorptive capacity* or to overcoming problems of low absorptive capacity of current national research system, which is a fundamental component of the National Innovation System. While this dimension has not been investigated deep enough, this author could conclude, like Nerdrum and

Gulbrandsen (2007), that it occurs through two mechanisms: one is through the numerous projects and personal contacts established between industry and the centre's staff and the other through the relatively high staff turnover which they suffer from and which benefits industry, consultancy offices etc. Third, ITCs have a *lubrication* or *intermediation* role in the research and innovation system. They have close relations to higher education, namely through joint laboratories and joint projects. This gives them the possibility to act as a buffer zone between universities and industry with the beneficial effect this produces, and also play an intermediation role between the two. They can therefore help SMEs of the private sector in particular increase their capacity for R&D by acting as a *flexible repository* in the national innovation system.

In Europe, industrial technical centres are found in France, Belgium and Spain. France plays a leadership role in this respect and has been particularly active in terms of the number of patent applications. It was ranked fourth in the world in 2006 and fifth in the world in terms of the size of R&D spending - behind USA, Japan, China and Germany. In all, 17245 patents were nationally registered in 2006 mainly by institutions - on top of individual applications. An average of 80 000 patent applications is lodged each year at the INPI (Institut National de la propriété Intellectuelle). Part of these applications come from ITCs. This requires anticipating industrial needs, the conduct of collective R&D projects, the diffusion of technological know-how and the transfer of R&D results to industry.

Most of the ITCs were created between 1948 and 1964[3] and their original mission was to provide support to the industrial sector, made up to a large extent by SMEs. Their legal framework was recently integrated into the Research Law, putting thus emphasis on their R&D function[4]. They constitute the first Applied Research Network in France and have been providing for the last fifty years a set of expertise to SMEs which they would not otherwise acquire individually in the fields of R&D. In their role as an intermediary institution, between the world of research and that of enterprises, they play the role of facilitators, the exchange of information, the acquisition of know-how and the diffusion of technological progress and technology transfer. ITCs represent nowadays 4500 collaborators, 50% of whom are engineers, scientists, and managers,

Intensive exchange occurs between ITCs and both industrial entities and universities. Thus annually, 200 theses are passed and 50 post-doctoral studies are supported on top of 200 contracts with public laboratories.

ITCs appear to have several attractions to Maghreb countries and indeed to LDCs as a whole (Lachat, 2004). Firstly, they are heavily based on incremental innovation and applied research, which in the case of innovation take off, represent the bulk of innovation needs. Breakthrough and radical innovation appear at this stage non feasible. Secondly, innovation take off is the most delicate task when it comes to SMEs of the private sector. In other words, experience has shown that ITCs are particularly suited for R&D and innovation for the small scale industry. Thirdly, innovation take off has to occur often in situations where NSIs are incomplete, unstructured and yet to take full shape (Djeflat, 2003). ITCs appear to be particularly suited for this type of NSI. Fourthly, ITCs cater, as seen earlier, for collective R&D and innovation needs of small enterprises. They can thus constitute a relatively good substitute for the lack of capabilities and means that SMEs suffer from particularly in LDCs and specifically in the Maghreb countries.

5. Concluding Remarks

In this paper, an attempt is made to set the basis for the issue of inno-vation take off both as a concept and as a policy instrument. This is in response to the issue of catch up, which concerns a limited number of countries namely the emerging ones. In a situation of de-structured, fragmented, immature and sometimes virtually non-existent NSI, in which most African countries find themselves, it would be difficult to work out catch up strategies. Several attempts are being made to build innovation systems, inspired by those in advanced countries. These attempts are centralized and state-supported and take very little account of the creative potential of these countries, which are largely decentral-ized in nature, hence the very limited results obtained. This is an indication that innovation strategies built on the catch up paradigm may not be totally appropriate in the current stage of development that many African countries are.

The take off paradigm proposed in this paper takes into account this largely decentralized creative potential and the effective capacity to mobilize knowledge resources, and to exercise a sufficiently powerful thrust for innovation to take place through the emergence of effective demand for R&D. This approach incorporates implicitly the gathering of the necessary conditions for catch up at a later stage. The stages for innovation take off need to be further examined both theoretically and from an empirical point of view. They need to take into consideration the specificities of countries, knowing that take off is context specific and strongly localized, even if common elements such as effective demand, public procurement and guarantee to the innovation effort exist. In African countries, the weakness of the private sector and market imperfections place responsibility on public procurement to get innovation systems to take off.

Decentralized mechanisms such as industrial technical centres could constitute an appropriate mechanism to help the state action in its take off strategy, even if they suffer from a large deficit in terms of academic research. They have many interesting features of NSI in the construction stage and could have a structuring effect on the rest of the innovation system. This is a research avenue to be further explored.

Notes

[1] Jay-Ik Choi, Vice President of the Korean Industrial and Technological Association, the R&D activities in the Korean enterprises, ANRT, Paris, March 2004.

[2] Extract from the White Book on ITCs, the ITCs Network, 2005, (France).

[3] Law of 22 July 1948.

[4] Research (Art. L. 342-1 à L. 342-13).

References

Ait Habouche, M. Jaidi, L. and Zaidi, D. (2004), *Gouvernance et institutions intermédiaires dans les processus d'innovation,* Femise Report, Université de la Méditerrannée, Marseille, France.

Archibugi D., Howells J., and Michie J. (1998), 'Innovation systems in a global economy,' *CRIC discussion Paper*, No.18, Centre for Research on Innovation and Competition (CRIC), The University of Manchester, UK.

Arocena R. and Sutz J. (1999), 'Looking at national systems of innovation from the South,' in *Industry and Innovation*, 7(1): 55-75.

Arocena R. and Sutz J. (2000), 'Interactive learning spaces and development policies in Latin America,' *DRUID Working Paper*, 13/2000, DRUID, Copenhagen Business School, Denmark.

Arocena R. and Sutz J. (2002), 'Innovation Systems and Developing countries,' *DRUID Working Paper*, No. 02-05, DRUID, Copenhagen Business School, Denmark.

Arocena R. and Sutz J. (2003), 'Understanding underdevelopment today: news perspectives on NSI,' *First Globelics Conference: Innovation Systems and Development Strategies for the Third Millennium*, Rio de Janeiro, Brazil.

Avnimelech, G. and Teubal, M. (2006), 'Innovation and Technology Policy for Catching Up: A Three Phase Life Cycle Framework for Industrializing Economies,' *Studies and Perspectives Series*, No.69, United Nations, Economic Commission for Latin America and the Caribbean.

Carré H. (2002), 'Innovation et développement dans une économie de rente: le cas du secteur agro-alimentaire au Sénégal,' in Djeflat A. and Boidin B. (eds.), *Ajustement et technologie en Afrique*, Paris: Publisud.

Cohen W. and Levinthal D. (1989), 'Innovation and learning: the two faces of R&D,' in *Economic Journal*, 99(397):569-596.

Casadella V. (2006), *Le Système de Construction de Compétences au Sénégal*, Thesis de doctorat, Université de Perpignan, Perpignan, France.

Cassiolato J.E and Lastres H.M. (1999), 'Local, national and regional systems of innovation in the Mercosur,' *DRUID Conference paper*, DRUID, Copenhagen Business School, Denmark.

Devalan P. (2006), *L'Innovation de rupture : Clé de la compétitivité*, Hermes-Science-Lavoisier, Paris, France.

Djeflat A. (2003), 'Les Systèmes Nationaux d'Innovation: entre globalisation et territorialisation,' in Michel Rautenberg (ed.), *Dynamiques locales et Mondialisation*, Revue CLES, l'Harmattan, Octobre, pp. 131-153.

Djeflat A. and Boidin B. (eds.) (2002), *Ajustement et Technologie en Afrique*. Publisud Paris, France.

Djeflat A. (2004), 'National Systems of Innovation in the Mena Region,' *World Bank Institute Report*, Washington DC.: World Bank.

Djeflat A. (2006), 'Le système de support technologique (SST) au Maghreb: cas des PME en Algérie et en Tunisie,' in H. Khelfaoui (ed.), *L'intégration de la science au développement, expériences maghrébines*. Publisud Paris, France.

Djeflat A., Devalan P., and Youcef Ettoumi F. (2007), 'Evaluation des Politiques et Programmes d'innovation dans le secteur industriel,' *Final Report*, European Commission - Ministry of Industry, Brussels: EC.

Djeflat A. (2008), 'Innovation takes off through Industrial Technical Centers in Maghreb Countries: a missing link in NSI or new opportunity?,' *Sixth Globelics International Conference*, Mexico City, Mexico, 22-24 September.

Edquist C. (1997), Systems of Innovation, Technologies, Institutions and Organizations. London: Pinter.

Edquist, C., Hommen, L. and Tsipouri, L. (eds.) (2000), *Public Technology Procurement and Innovation*, Boston, USA: Kluwer Academic Publishers.

Edquist C. (2001), *System of Innovation for Development*, UNIDO World Industrial Development Report (WIRD), Vienna: UNIDO.

Economic Commission for Africa (2004), Renforcer la compétitivité des petites et moyennes entreprises africaines, un cadre stratégique d'appui institutionnel, ECA/DMD/PDS/TP/UNU. Addis Ababa: United Nations.

Fagerberg J., Guerrieri P., and Verspagen B. (eds.) (1999), *The Economic Challenge for Europe:Adapting to Innovation Based Growth*, Cheltenham, UK: Edward Elgar.

Fagerberg, J. and Godinho M. (2004), 'Innovation and Catching-up,' in Fagerberg J., Mowery D., and Nelson R., *The Oxford Handbook of Innovation*, Oxford, UK: Oxford University Press.

Foss N.J. (1999), 'Networks, Capabilities, and Competitive Advantage,' in *Scandinavian Journal of Management*, 15(1):1-15.

Gu S. (1999), 'Concepts and methods of NIS approach in the context of less-developed economies,' *DRUID Conference Paper*, DRUID, Copenhagen Business School, Denmark.

Hobday M. (1995), *Innovation in East Asia: The Challenge to Japan*. Cheltenham, UK: Edward Elgar.

Hobday M., Rush H., and Bessant, J. (2004), 'Approaching the innovation frontier in Korea: the transition phase to leadership,' *Research Policy*, 33(10):1433-57.

Johnson B. (1992), 'Institutional learning,' in Lundvall B.-Å., *National Innovation System: Towards a theory of innovation and interactive learning*, London: Pinter Publishers.

Johnson B. and Lundvall B.-Å., (2003), 'National System of Innovation and Economic development,' in Muchie M., Gammeltoft P., and Lundvall B.A. (eds.), *Putting Africa First: The Making of African Innovation System*, Aalborg, Denmark: Aalborg University Press.

Kim L. (1997), Imitation to Innovation: The dynamics of Korea's Technological Learning, Boston MA, USA: Harvard University Press.

Kirat T. and Lung Y. (1999), 'Innovation and proximity – territories as loci of collective learning processes,' in *European Urban and Regional Studies*, 6 (1): 27-38.

Korean Intellectual Property Office, cited by Yim D. S. (2004), *Korea's National Innovation System and the Science and Technology Policy*, Science and Technology Policy Institute (STEPI). See: http://www.unesco.org/science/psd/thm_innov/forums/korea.pdf (Accessed: July 2009).

Lachat J.C. (2004), 'Accompagnement au processus de modernisation du Ministère de l'Industrie et des Organismes liés Expertise d'appui à l'Industrie: Centres Techniques,' *Rapport pour le compte du Ministère de l'Industrie*, novembre.

Lall S. (2000), Technological Change and Industrialization in the Asian Newly Industrializing Economies: Achievements and Challenges, in Kim L. and Nelson R., Technology, Learning and Innovation, Cambridge: Cambridge University Press.

Lall S. and Pietrobelli C. (2002), Failing to Compete: Technology Development and Technology Systems in Africa. Cheltenham, UK: Edward Elgar.

Lall S. and Pietrobelli C. (2003), 'Manufacturing in sub-Saharan Africa and the need of a national technology system,' in Muchie M., Gammeltoft P., and Lundvall B.A. (eds.), *Putting Africa First: The Making of African Innovation System*, Aalborg, Denmark: Aalborg University Press.

Liu X. and White S. (2001), 'Comparing Innovation Systems: a framework and application to China's transitional context,' *Research Policy*, 30(6):1091-1114.

Lundvall B.-Å., (1992), National Systems of Innovations. Towards a theory of innovation and interactive learning, London: Pinter Publishers.

Lundvall B.-Å., Johnson B., Andersen E.S, and Dalum B. (2002), 'National systems of production, innovation and competence building,' *Research Policy*, 31(2):213-231.

Malerba F. (2004), Sectoral Systems of Innovation: Concepts, issues and analyses in six major sectors in Europe, Cambridge: Cambridge University Press.

Masinda M. (1998), National Systems of Innovation: Implications of Science and Technology Policies in Sub-Saharan Africa, CPROST Report.

Maskell P. (2001), 'Towards a knowledge-based theory of the geographical cluster,' in *Industrial and Corporate Change*, 10 (4), 921-943.

Ministry of Science and Technology & Korea Institute of Science and Technology Evaluation and Planning (2004), cited by Jay-Ik Choi, Vice President of the Korean Industrial and technological Association: les activités de R&D dans les entreprises Coréennes, ANRT Répères sur l'innovatin en Corée, Paris (March).

Mota J. and De Castro L.M. (2003), 'Connecting Capabilities through Technological Centres,' *Paper Prepared for the 19th IMP Conference*, University of Lugano, Switzerland, 4th – 6th September.

Mowery D.C. and Oxley J. (1997), 'Inward technology transfer and competitiveness: the role of national innovation systems,' in Archibugi D. and Michie J. (eds.), *Technology, globalization and economic performance*, Cambridge: Cambridge University Press.

Muchie M. (2003), 'Re-thinking Africa's development through the National Innovation System,' in Muchie M., Gammeltoft P., and Lundvall B.A. (eds.), *Putting Africa First: The Making of African Innovation System*, Aalborg, Denmark: Aalborg University Press.

Mytelka L. (2000), 'Local systems of innovation in a globalized world economy,' in *Industry and innovation*, 7(1):15-32.

Naclerio A. (2004), La dimension systémique du Système National d'Innovation: Une application au cas de l'Argentine. Thèse de doctorat, Université Paris 13.

National Syndicate of Researchers (2006), cited by H. Khelfaoui *Les technologies de l'information , politique de recherche et d'innovation et politique scientifique*, Report to the Ministry of Urban Planning and Environment, Algeria, p.12.

Narula R. (2004), 'Understanding absorptive capacities in an Innovation Systems context: Consequences for Economic and Employment Growth,' *DRUID Working Paper*, No. 04-02, DRUID, Copenhagen Business School, Denmark.

Nerdrum, L. & Gulbrandsen, M. (2007), 'The technical-industrial research institutes in the Norwegian innovation system,' *TIK Working paper on Innovation Studies No. 20070614*, NIFU STEP, Oslo, 31 pages.

Nielsen R.N. (2005), 'Does university education contributes to human resources development in the Danish system of innovation and competence building?,' *DRUID Conference Paper*, DRUID, Copenhagen Business School, Denmark.

Niosi J., Bellon B., Saviotti P., and Crow M. (1992), 'Les systèmes nationaux d'innovation: vers un concept utilisable,' *Revue française d'économie*, Paris, 7(1): 215-249.

Oyelaran-Oyeyinka, B. (2004), 'Learning and local knowledge institutions in African industry,' *UNU/INTECH Discussion Paper Series*, Maastricht, Netherlands.

Rostow W.W. (1960), *The Stages of Economic Growth: A Non-Communist Manifesto*, Cambridge: Cambridge University Press.

Szogs A. (2004), 'The making of innovation systems in least developed countries. Evidence from Tanzania,' *DRUID Conference Paper*, DRUID, Copenhagen Business School, Denmark.

Viotti E. (2002), 'National learning systems. A new approach on technological change in late industrializing economies and evidences from the cases of Brazil and, South Korea,' *Technological Forecasting and Social Change*, 69(7): 653- 680.

Viotti E. (2003), 'Technological Learning Systems, Competitiveness and Development,' *First Globelics Conference: Innovation Systems and Development Strategies for the Third Millennium*, Rio de Janeiro, Brazil.

Von Hippel E. (1988), *The Sources of Innovations*, Oxford: Oxford University Press.

Wangwe S. M. (2003), 'African systems of innovation: toward an interpretation of the development experience,' in Muchie M., Gammeltoft P., and Lundvall B.A. (eds.), *Putting Africa First: The Making of African Innovation System*, Aalborg, Denmark: Aalborg University Press.

African Journal of Science, Technology, Innovation and Development
Vol. 1, No. 1, 2009
pp. 76-102

A Firm-Level Analysis of Technological Externality of Foreign Direct Investment in South Africa

Abdelrasaq Al-Suyuti Na-Allah* and Mammo Muchie**

Abstract

This paper examines how Foreign Direct Investment (FDI) impacts on productivity performance in host economy by raising and addressing two key questions: (i) To what extent does technological externality of FDI (FDI Spillover) rely on host economy's skill deficiency characteristic before it can be generated and appropriated? (ii) Is there a regional dimension to the level of spillovers from multinationals to domestic firms? Using establishment level data for the South African economy, we estimate regression equations for a representative sample of her manufacturing plants. Our findings reveal that regardless of the environmentally imposed skill deficit factor, foreign firms are able to generate productivity spillovers for their host. Successful diffusion to domestic firms is however mediated and significantly circumscribed by the skill factor. It also appears that some support exists for the claim that level of FDI spillovers to domestic firms relies on regional presence, i.e. closeness to the region in which foreign firms locate.

Keywords: Foreign Direct Investment Spillover, Technology Transfer, Skill Constraints, Agglomeration Economies.
JEL Classification: F21, F23, O24, R39

---

* SARCHI/NRF Research Fellow, Institute for Economics Research on Innovation, Tshwane University of Technology, Pretoria, South Africa. Email: abdelrasaq@yahoo.com

** DST/NRF Research Professor of Science, Technology, Innovation and Development, IERI, Tshwane University of Technology, Pretoria, South Africa; Professor, Aalborg University. Emails: mammo@ihis.aau.dk; MuchieM@tut.ac.za

1. Introduction

For much of the developing world, the decade of the 1980s was a turning point in experimental economics as market pessimism gave way to measured optimism in development policy engineering. The abandonment of import substitution strategies that were principally influenced by the dependency and Neo-Marxist theories of earlier decades saw governments across the South adopting many of the neoliberal policies of open and free market orthodoxy. As a major pillar of this new economic model, attracting foreign direct investment (henceforth, FDI) into the local economy became a policy priority for many governments. Consequently, varieties of generous investment incentive packages were developed by these governments in order to turn their economies into attractive destinations for investors.

For South Africa, this has been especially the case since the dismantling of Apartheid and subsequent formation of the democratic government in 1994. With her new found status as an integrated member of the global economy, the country joined the club of FDI contestants through its pronounced Growth Employment and Redistribution (GEAR) strategy in 1996. Several inducement packages were introduced at both national and provincial levels of governments to foster the growth of national stock of foreign capital. Besides being a signatory to the Trade Related Investment Measures (TRIMS) Agreement as well as over 30 bilateral investment treaties that have, as priorities, increasing the national stock of FDI, there were, as at 2004, more than 35 investment incentive schemes in place across the country (Gelb and Black, 2004).

A key motivation for these efforts by the South African authorities hinges around a widely held but controversially substantiated notion that there is a technological externality to FDI. Commonly referred to in the literature as FDI spillover, its proponents have argued that domestic economies in general and their firms in particular stand to reap techno-logical reward from the activities of foreign multinationals (MNCs) operating in their countries. Despite the intuitive appeal of this claim, empirical analyses to date have yielded mixed results (Crespo and Fontoura, 2007). To improve our understanding on what currently appears to be an impasse some analysts have called for more detailed investigations that involve for instance, adopting an analytical framework which incorporates spillover promoting/inhibiting specific national factors (Lipsey, 2002), as well as employing appropriate data, that is generally considered to be panel data with disaggregation at the firm level (Gorg and Greenway, 2001).

Past efforts at explaining the incidence of FDI spillover in South Africa have generally relied on aggregated country or industry level data to confirm its positive impact (Fedderke and Romm, 2006; Gelb and Black, 2004). However, it is now known that relying on results obtained with aggregated data at a macro as opposed to micro firm-level template may not provide an accurate understanding of FDI gains, especially if the observed positive impact at national or industry level also comes with negative distributional implications. For instance, it is quite possible that by virtue of location specific attributes such as agglomeration, productivity gains attributable to FDI at a more aggregative macro level only accrue to a subset of firms within the entire national population. Or even if all firms were to benefit, chances are also there that these benefits will not be evenly spread across board. A scenario of this sort presents a distributional distortion that in the views of Crespo and Fontoura (2007) calls for research attention. In articulating their argument the authors made the following remark:

> ...although FDI may work as a convergence mechanism at the national level, if it produces significant gains in efficiency for domestic firms, it can also increase domestic inequalities at a regional level. This is a result that certainly justifies further research (Crespo and Fontoura, 2007: 417).

Another limitation with current evidence from South Africa is that to date there has been no visible attempt at accounting for factors that promote or inhibit spillover occurrence especially with respect to the role of environmental specificities that Lipsey recommended. Yet it is well known also that the South African economy, like those of other developing ones, is encumbered with many deficient characteristics, among which skilled manpower shortage occupies a pride of place. Indeed, as recently as 2009, a report carried on the website of Grant Thornton cited International Business Report as finding for the third consecutive time that skill shortage remains the greatest constraint to business growth in South Africa (Grant Thornton, 2009). There are also several Survey Reports produced by both local and international organizations that have documented the acuteness of manpower shortage in the economy[1]. As we shall see later in this paper, if theoretical predictions of the implications of skill shortage for FDI spillover are anything to go by, an investigation of the impact of foreign capital on productivity performances of South African firms would surely be a revealing exercise.

In the light of all these considerations, we are inclined to define research questions to be investigated in ways which depart from previous attempts as follows: (i) To what extent does FDI spillover rely on host economy's skill deficiency characteristic before it can be generated and appropriated? (ii) Is there a regional dimension to the level of spillovers from multinationals to domestic firms? To address these issues, we estimate regression equations for a representative sample of South African manufacturing firms using information obtained from World Bank Enterprise Survey database. Our findings reveal that regardless of the presence of this environmentally imposed constraint, foreign firms are able to generate technological spillover for their host. Successful diffusion to domestic firms is however, mediated and severely circumscribed by the skill deficiency factor. Furthermore, support for the claim that level of FDI spillovers that diffuses to domestic plants relies on regional presence is weakly revealed in our evidence.

This paper is organized in six sections. We start with a review of relevant literature in Section 2, followed by motivation and specification of all hypothesised relationships in Section 3. In Section 4 relevant issues around model specification are treated alongside data and variable construction before presenting results in Section 5. The final section summarises and presents the paper's concluding remarks.

2. Skill Constraints, Agglomeration Economies and Technological Externality of FDI Debate

The Basic Argument: Despite its recognition in early attempts to explore the impact of FDI on host country's economic performance, the productivity spillover dimension did not receive a formalised comprehensive theoretical treatment until the late 1970s when contributions from experts like Findlay (1978), Koizumi and Kopecky (1977) and Das (1987) among others began to appear in the literature[2]. In their various quests to explain the complex mechanism that generates the so-called "positive contagion" effect of foreign capital on host productivity, a very plausible but refutable proposition was made.

Assume as it is argued to be the case that foreign capitals of MNC subsidiaries embody some firm specific assets otherwise called technological advantages. Being backward offers the domestic firms in the FDI importing country an advantage to bring own technology level to those of their resident foreigners as long as they can capitalise on the now available opportunity to copy or learn from the advanced technology of MNCs. The importance of relative backwardness is very much emphasised as a key determinant of potentials to close the technology gap. Thus the lower the host's technology level relative to the foreigner's, the better the chances and magnitude of successful technology diffusion. In the final analysis, the model has prediction to the effect that the larger the host's stock of foreign capital relative to domestic ones the higher the rate of her technical progress.

After nearly four decades of empirical research, it is hard to find concrete evidence in support of this prediction. In fact available evidence seems to weigh heavily against it. In a fairly recent survey of the literature, Gorg and Greenway (2001) reported that most works failed to find positive spillover impact of FDI while evidences of negative spillover were even discovered in a good number of studies. With regards to studies in the survey that used firm-level panel data which is generally considered to be the appropriate technique of inquiry of this kind, only 2 out of a total of 16 found significant positive spillover effect. Five of these studies found negative impact while the rest could not establish any statistically significant relationship between FDI and domestic firms' productivities.

Local Environmental Conditions, Skill Shortages and Agglomeration: Attempts to make sense of these divergent results have seen critics launch a two-pronged attack on both theoretical and empirical fronts. Those who grudge the theory argue that technological diffusion does not only depend on relative backwardness of FDI importer but also on a number of other host's economic characteristics. In what can be described as a non mainstream explanation Marin and Bell (2004) argue that certain aspects of the local environment that create constraints and opportunities for MNC subsidiaries do matter for FDI spillover occurrence. In their unconventional exposition, the authors present a framework which conceptualises a typical MNC subsidiary as a knowledge creating entity and not, according to conventional model, a passive knowledge transferring intermediary that simply receives knowledge initially created outside by its parent company and delivers it to beneficiaries in the host economy. In this view, if local environmental conditions like socio-political system, labour markets, skill structures and infrastructures are not supportive of the knowledge creation process, no spillover will be created let alone appropriated. A particularly striking aspect of their contribution lies in the message that it is what the MNC subsidiaries actually do in the host country as may be revealed in knowledge-creating and knowledge-accumulating activities as well as embeddedness in linkages with domestic firms that matter for spillover. Since these activities are essentially independently driven at local levels and not dictated or directed from headquarters we are reminded by the authors that the mere existence of MNC subsidiaries is not what is really important for technology to be generated and its value appropriated.

But while this brand of argument relates to the supply-side of spillover another paradigm as revealed in works on absorptive capacity focuses on demand side with similar conclusion that local environmental conditions like those mentioned above must be right before there can be incidence of spillover (Wang and Blomstrom 1992; Perez, 1997; Kinoshita, 2001; Hermes and Lensink, 2003). Also, Baskaran and Muchie (2009) suggest that weak national system of innovation is linked to low absorptive capability of technology spillover from FDI. They established possible relationships between the characteristics of a National System of Innovation (NSI) and their impact on FDI outcomes, particularly in developing countries. They employed a heuristic NSI-FDI conceptual framework linking the robustness of NSI to the benefits or lack of it from FDI and applied this to analyse descriptive data from selected developing economies – China, India, South Africa, Ghana, Ethiopia, Tanzania and Zambia (Baskaran and Muchie, 2009; Muchie and Baskaran, 2009).

The concept of absorptive capacity as argued in many of these contributions actually presupposes that spillover does not accrue in a vacuum. Rather, it is influenced by certain capacity related characteristics of the host. This capacity can be thought of in terms of a required threshold in the level of development (Xu, 2000), especially as it relates to the skill quality of labour force (Blomstrom *et al.*, 1994; Kokko and Blomstrom, 1995). Whereas the requirement of relative backwardness presupposes the existence of some technology gaps between the MNCs and their hosts before spillover can materialise, the argument that a threshold in level of labour force's skill is needed suggests that this gap should not be too wide. When it is too wide, as may be captured in say, relatively highly unskilled (or acute shortage of skilled) labour force, the prospect of the host's firms engaging moderately skilled workforce with capacity to copy and adapt the technology of MNCs for their own purpose will be impaired. Conversely, a relatively moderate gap with sufficient assurance that the host's workforce is in possession of the right quantum of skill needed to internalise the foreigners' best practices will ensure successful diffusion. This argument would lead us logically to view differences in skill level of workforce or national absorptive capacities in general as one possible explanation for the divergent results that are revealed in the literature.

In the context of South Africa therefore, one may be tempted to suspect that her problem of inadequate supply of skill manpower would definitely constitute an inhibiting force to the realisation of FDI spillover in the economy. But while this may be a significant problem in its own right there are also other environmental factors that either place significant demand on skill requirement as a key factor in FDI technology diffusion or raise the stake for investing in its supply. On the demand side, the capital intensive nature of FDI that the country has attracted over the past years places relatively enormous demand on highly skilled workforce before knowledge can be generated and transferred. As Figure 3 reveals for instance, the distribution of foreign firms is positively skewed in favour of the more capital intensive and skill demanding sectors like Motor Vehicle, Chemical & Allied Products, Electrical & Electronics and Machinery. On the supply side, it can be argued that the prevalence of such national calamities as crime and AIDS creates uncertainties and prevents medium and long term commitment of firms by discouraging investment in training and development of staff. Despite her problem of skill shortage a recent World Bank Report reveals that relatively few firms in the country, in comparison with other countries like China and Brazil, have training programmes for their staff. About 45 per cent of firms interviewed had expressed concern over the impact of the AIDS epidemic and that this concern had a strong impact on their investment (Clarke *et al.*, 2005).

But recent development in the of analysis of skill factor as an element of absorptive capacity has shifted the focus beyond national level accounting to that of a sub-regional one and in the process draws attention to the possibility of a regional bias in FDI spillover effect. It is being claimed that its occurrence may vary among the constituent regions of a nation depending on the skill characteristics of regional workforces. The implied hypothesis that we may expect to associate regions with high level of educational attainment or sufficient presence of skilled workforce with successful FDI technology adoption and those with low level or insufficient presence with limited success at best has been confirmed in some studies (Ponomareva, 2000; Yudaeva *et al.*, 2003).

Reinforcing the above allusion to regional distortion in spillover effect is the argument of those who model economic geography as a determinant of technology diffusion. According to this view, there are several reasons to expect that all regional constituents of a nation will not benefit in equal measures from the activities of MNCs. Some of these include: (i) both demonstration effects and mobility of labour from foreign to domestic firms which constitute two key channels of technology transfer may be confined to local areas where MNCs are located; (ii) due to transport costs, vertical linkage that is also another source of technology diffusion may be bounded in space. Since all these factors operate in favour of those native firms with significant presence in or around the regions where MNCs are present the expectation follows that much of the technology that will diffuse to domestic firms will be confined to these regions. Several authors have pushed this argument beyond mere rhetoric by empirically showing that FDI spillover has a circumscribed regional dimension (Renato, *et al.*, 2007; Torlak, 2004; Girma, 2003; Girma and Wakelin, 2001).

With respect to the empirical evidence, the major issues that have arisen are centred on appropriateness of research design employed in many of the studies. For instance, the use of aggregate country or industry level data which appears to be popular with spillover empirical studies rather than firm level ones does not allow for detailed investigation by controlling for other factors (Gorg and Strobl, 2001). As further pointed out by Gorg and Greenway (2001), the notion of building empirical conclusions around cross-sectional as opposed to panel data evidence that is the case with many of the existing investigations does not allow the time variant nature of the studies to be captured. A final aspect of empirical inadequacy is implied in Lipsey's recent suggestion that a detailed analysis of the different circumstances and policies of countries, industries and firms that promote or obstruct spillovers is needed for better understanding of how FDI impacts on productivity (Lipsey, 2002).

South Africa-based Studies: A literature survey of FDI spillover studies that have so far been conducted for the South African economy reveals that the emerging body of evidence still has much to learn from many aspects of the critics' arguments mentioned above. For instance, in Fedderke and Romm (2006), spillover was modelled in a growth accounting framework using aggregated data at national level to confirm its positive impact. Gelb and Black (2004) represents another attempt at accounting for the incidence of spillover by using statistical averages at sector level to shed some lights on MNC subsidiaries' expenditure on training and human capital development in South Africa. As these studies neither accounted for the role of environmental specificities such as skill constraint factor nor employed firm level data as we did in this paper, the departure and relevance of our study has hopefully been well established.

3. Main Hypotheses

All the hypotheses investigated are inspired by the basic theoretical proposition that foreign capital generates technological returns for the host economy and empirical convention which captures this argument in terms of some qualified pattern of relationship between proxies for foreign presence and productivity performances in the economy.

Following Aitken and Harris (1999), the productivity spillover of FDI can be measured at both plant and industry levels. In the former case, if foreign capital embodies advanced technology we should, other things being equal, associate establishments with foreign presence with significant productivity gains. But all things are not equal in a skill constrained economy. It is arguable that as a result of skill shortage, enterprises' chances of engaging employees with the right kind of skills needed to internalise the foreign technology will be constrained. However, we argue here that inability to attract skilled workforce will hardly constitute a significant problem for the MNCs. Because of their better incentives, they can price out their domestic counterparts from the marginal skilled labour market that exists thereby equipping own plants with capable hands[3]. We have come to know as well that when skill shortage is the problem, MNCs do provide training for employees in order to raise their level of competence to that required by them. Hence, our first hypothesis will not depart from conventional wisdom which says:

> Hypothesis 1: Plant level foreign presence is positively related to productivity performance and those with more foreign equity participation should exhibit higher productivity performances than their comparable domestic counterparts.

In the latter case, economists have proffered several reasons why the presence of foreign technology in an industry should have a positive contagion effect on domestic technology of the industry. First, the demonstration effect makes it possible for domestic plants to imitate or copy the technological practices of their foreign competitors (Wang and Blomstrom, 1992; Barrios and Strobl, 2002). Second, workers from MNC subsidiaries can move to local companies and in effect deploy the skill and knowledge they acquired in the former for the latter's benefit (Glass and Saggi, 2002). Third, vertical linkages by foreign firms with local suppliers or customers also afford the natives the opportunity to learn the best practices.

All these various channels of technological leakages have implicit assumptions that learning, on the part of domestic firms' employees, is concurrently and effortlessly taking place. But this may not be so if by virtue of skill shortage in the economy these plants are unable to engage the right calibre of workers. This is likely to be so especially if the MNCs completely absorb all of the marginal skill manpower that exists in the economy and jealously guard them from leaving to work for domestic competitors. Employees with inadequate skills will almost certainly have difficulty copying or imitating the advanced technological practices of MNCs. In similar manner domestic suppliers/customers with poorly trained staff who interact with foreign companies may practically learn little or nothing from their interaction[4]. All these blur the chances that technology will get transmitted through the previously discussed channels. Hence the question of how well externalised foreign knowledge gets diffused to the locals is now a question of how well the locals are prepared to make the right kind of investment in human capital in order to increase their absorptive capacity. Those who invest well enough are more likely to do a better job attracting capable hands and absorbing foreign technology than their counterparts, other things being equal. Hence the second hypothesis for investigation can be stated thus:

> Hypothesis 2: Domestic firms with relatively highly skilled workforce are more successful in absorbing technology spillover from resident MNC subsidiaries.

Finally, it is understandable to have some people argue that levels of technology that do spillover to domestic establishments have a regional tone. This is simply because by virtue of posession of certain competitive attributes some regions are relatively better at attracting manufacturing activities to their areas. These attributes as may be revealed in the presence of better economic infrastructures or as the New Economic Geography literature would have us believe, access to certain external economies confer some significant economic advantages on the locals. For instance, being close to regional clusters may facilitate access to common pool of skilled labour or enhance knowledge linkages among management staff. Linkages with suppliers/distributors are also other sources of external economies that come with agglomeration. If foreign firms are driven to locate in this type of region, as it is likely, then for the simple reason that proximity enhances the effectiveness of technology diffusion channels we may expect to observe domestic firms in the region absorbing more of the spillover returns than their counterparts elsewhere[5]. For instance, effective capture of the diffusion benefits of demonstrating a new product or production process favours those that are closer to the place where such demonstration takes place. Also, if a highly skilled MNC employee decides to leave and transfer the knowledge he has gained to domestic firms it is the ones located nearby that stand the better chance of engaging his service. Furthermore, linkages with local suppliers and distributors may be limited to the region because of transportation cost. The hypothesis therefore is that:

> Hypothesis 3: The level of spillover absorbed by domestic establishments is greater for those located in regions with more foreign presence.

4. Model Specification, Data and Variable Construction

The Model: The validity or otherwise of these hypotheses were tested by specifying and implementing an augmented Cobb–Douglas production function incorporating spillover and other productivity enhancing variables for a sample of South African manufacturing firms. We also included as controls other environmental constraints that appear to be of equal importance as skill shortage. Generally, the model estimated is of the form:

$$Q_{ij} = C + ß_1X_{ij} + ß_2Pe_{ij} + ß_3Ec_{ij} + ß_4fpp_{ij} + ß_5fpind_j + ß_6aggloj + e_{ij} \ldots\ldots\ldots 1$$

where:

Q_{ij} = output for plant i in industry j

X_{ij} = vector of production input for firm i in industry j

Pe_{ij} = vector of other productivity enhancers that include human capital/skilled labour

Ec_{ij} = measure of other environmental constraints

fpp_{ij} = measure of foreign presence at plant level

$fpind_j$ = measure of foreign presence at industry level

$ß_6agglo_{jl}$ = measure of foreign presence at industry level in regional location.

e_{ij} = usual error term

For impact on domestic firms' estimations both dependent and control variables only reflect data for domestic firms as required by hypotheses 2 and 3.

In the above specification, output is assumed to be a function of not only measures of input variables but also different measures of spillover, environmental constraints, agglomeration and other productivity influencing factors. The coefficients of variables of interest (spillover) shall be interpreted as evidence of the impact of foreign presence, which may confirm or refute our hypotheses. Data for the equation (Eq. 1) which represents the basic model for estimating several regressions in accordance with the dictates of our hypotheses have been expressed in the log-liner form and Ordinary Least Square technique is employed all through. The data and preliminary evidence arising from casual analyses that were performed are discussed next.

Data and Preliminary Evidence: Data for the investigation come primarily from The World Bank Survey of Enterprise in South Africa. The said Survey was conducted and completed for about 800 establishments between January and December 2004. A stratified random sampling methodology that generated 603 manufacturing establishments from the 800 that were sampled was followed. The framework of the said sampling technique was carefully designed to generate large enough sample size for a given industry. Level of precision with respect to both estimates of population proportion and mean of log of sales at industry level was fixed at 7.5 per cent for 90 per cent confidence interval. This simply suggests that except in 10 per cent of the cases there is a guarantee that the population parameter will be within the 7.5% range of the observed sample estimate[6].

The unit of analysis was the establishment defined as a factory, store or service outlet with own accounting identity and formally registered with the Department of Trade and Industry[7]. For the purpose of this research the words establishment, firm and plant will be used interchangeably to refer to this conception. The Survey which covers a three year period from 2000 to 2002 is cross-sectional meaning that we are constrained into using cross-sectional data in this analysis. Plants included in the exercise are mainly those with employment figures of 10 or more, although there are few cases of microenterprises where staff strengths are below 10 in number. The sectoral orientation of the manufacturing establishments that are the focus of this analysis corresponds to the ISIC revision 3.1 and only plants from 11 sectors that include Garment & Textile, Paper, Printing & Publishing, Food & Beverages, Rubber & Plastic, Electrical & Electronics, Wood & Carpentry, Chemical & Allied Products, Machinery, Furniture, Motor Vehicle and Other Manufacturing are taken for the investigation. After dealing with the problem of imperfect data we were left with 455 and 302 observations respectively for plant level and domestic level foreign presence estimations. Summary statistics with respect to variables used are provided in Table 1.

From the preliminary analysis of sampled evidence as revealed in the data we are able to confirm the validity of some of our underlining assumptions in this investigation. For instance, with respect to skills shortage, besides being the one environmental factor perceived by most South African manufacturing establishments as constituting the most serious obstacle to the smooth operation of their businesses, the problem is also found to have revealed the economy, as relatively uncompetitive among a comparator group of economies. A cursory look at Figures 1 and 2 below tells this story more. It is easy to observe from Figure 1 that more plants rated Workers' Skills as a Major or Very Severe Constraint to doing business in South Africa than any other investment climate constraint. Further, a multi-country comparison of plants identifying Labour Skill Level as a Major Constraint to their operation as given in Figure 2 also shows that nowhere except Brazil are more plants dissatisfied than South Africa.

It also appears that the two phenomena of capital intensity and plant productivity are positively associated with the incidence of foreign presence. In Figure 3 where distribution of foreign firms alongside sectoral line is given, it can be seen that two of the highly capital intensive sectors (Motor Vehicle and Chemical & Allied Products) reveal the highest levels of foreign penetration whereas the more labour intensive ones (Furniture and Wood & Carpentry Sectors) reveal some of the lowest levels. This is consistent with the finding in Fedderke and Romm (2006). Furthermore as contained in Figure 4 where the productivity dimension of this relationship is shown, it seems that there is a positive association between sector-level foreign presence and productivity: the three most productive sectors (Rubber & Plastic, Motor Vehicle and Chemical & Allied Products) are the sectors where foreign presence is most pronounced. On the other side, the lowest productivity performance is observed in the Furniture Sector with the least level of foreign penetration[8].

Figure 1: Percentage of South African Manufacturing Firms in 2002 rating Investment Climate Factors as Major or Very Severe Constraints

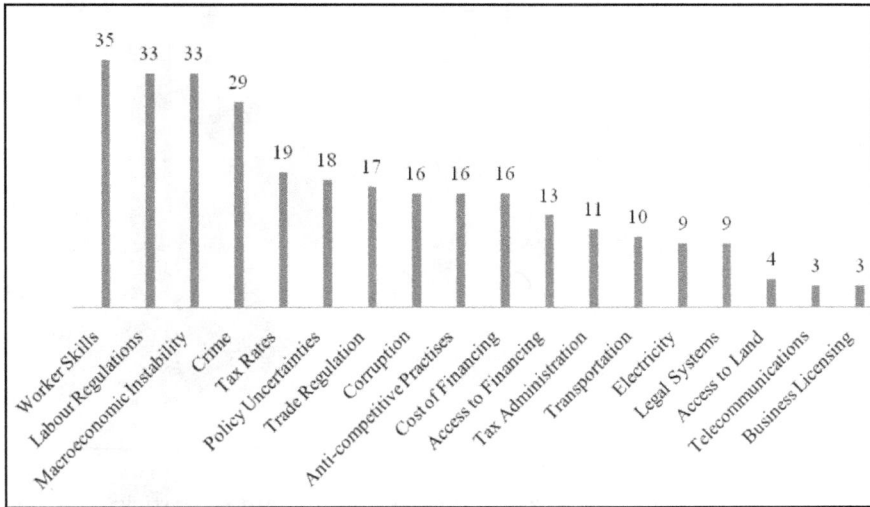

Source: World Bank Enterprise Surveys;
Note: Values are in Percentage

Figure 2: Percentage of Firms in each Country identifying Labour Skill Level as a Major Constraint

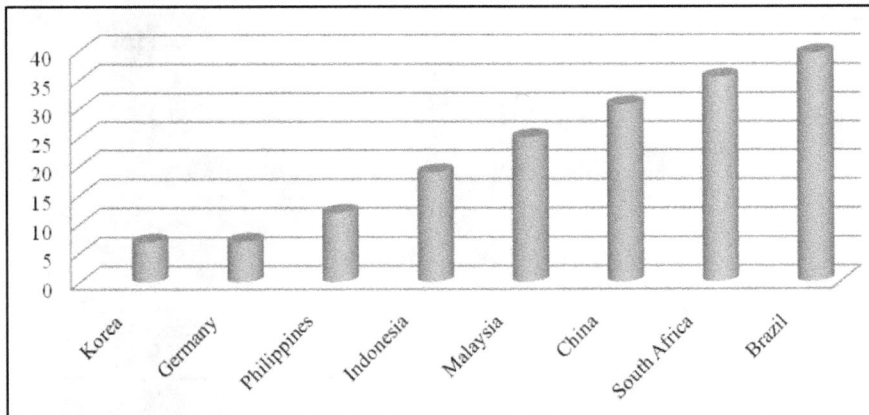

Source: See: http://www.enterprisesurveys.org/
Note: For all countries data are for 2003 except Germany and Korea (2005) as well as Malaysia (2002)

Figure 3: Percentage number of Foreign Firms in each Sector

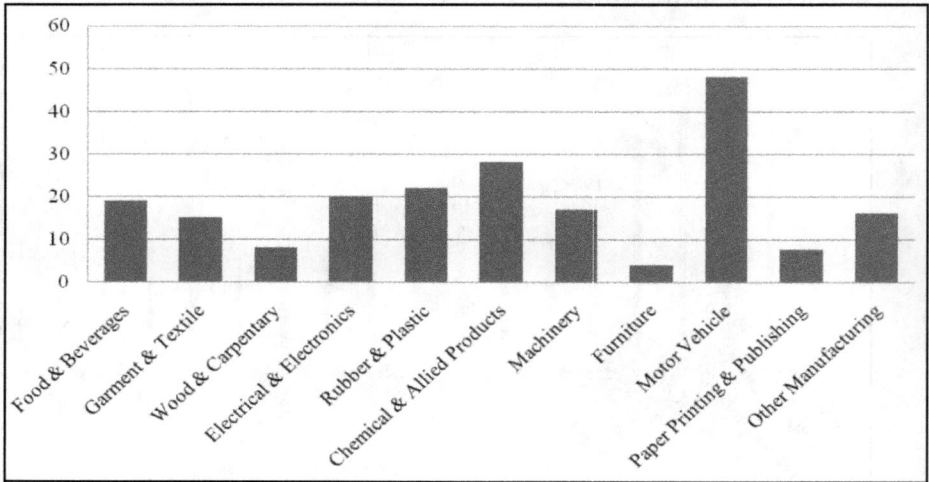

Note: Firms with ten percent or more foreign equity participation are classified as foreign

Figure 4: Average Productivity in each Sector

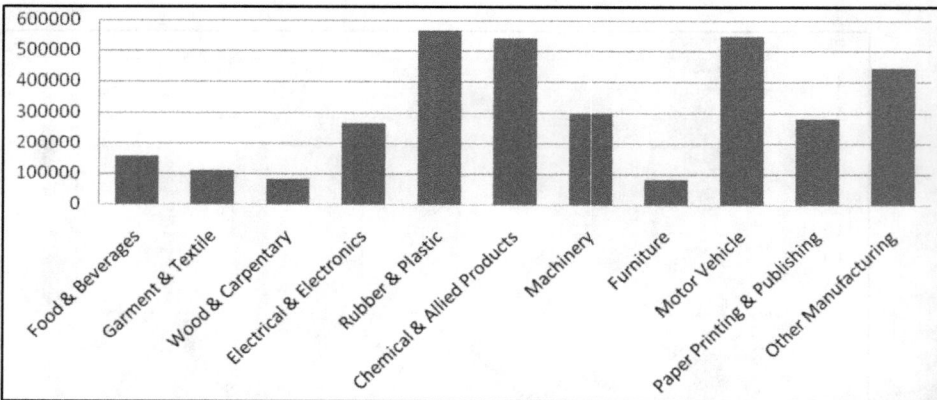

Note: Figures which are defined as market value of output divided by total number of workers are for the year 2002 only.

Variable Construction: Following standard practice the dependent variable is defined as the value of output. As measures of input factors, three variables have been included in the estimation. These are capital, labour and material. Capital is generated using information on plant fixed assets. It is calculated as the value of rent paid for machinery and equipment or value of depreciation incurred on owned assets. Labour is as usual defined in terms of the average number of workers (temporary and permanent).

Other productivity influencing variables that the literature suggests can be thought of as comprising various forms of knowledge factors that confer marketplace competitive advantage on firms over their rivals. Such knowledge can be acquired either through exposure gained as a result of exporting activities or through transferred technology from investment in plant and machinery. It could also become incumbent on an organization because such organization is driven by employees with more sophisticated skill background. The inclusion of this variable is very central to our analysis. These different sources have been captured by three different measures of productivity enhancing variables, viz, export intensity, investment in plant & machinery and human capital. Export intensity is defined as the percentage of establishment's sales exported directly. We measured investment in plant and machinery as the ratio of plant's new expenditure on machinery and equipment to output. Finally, proceeding from the logic that skilled workers attract relatively high wage compensation and that we should naturally expect the two events (skill and high wage bill) to coincide, human capital has been captured as the average wage bill of the firm. It is expressed as the ratio of a firm's wages to total number of employees. All these measures like measures of production input are expected to show positive associations with productivities.

The other environment specific constraints which appear to be of equal importance as skill in the data that were controlled for are: macro economic instability, labour regulation and crime. From the information provided in Figure 1 at least 29 per cent of firms interviewed rated each of these factors as Major Obstacle. Three variables captured in categorical terms *ecm, ecl,* and *ecc* proxy for the different influences of macroeconomic instability, labour regulation and crime respectively. A value of 1 was assigned if respondents considered any of the three factors a severe or very severe constraint and 0 otherwise. All these variables should be negatively related to productivity.

With respect to the spillover variables, foreign presence at plant level (*fpp*) is defined as the level of foreign participation or percentage of equity share capital owned by foreigners in a plant. The value varies between 0 and 100 per cent. Its coefficient is expected to turn up with a positive sign. The industry level version capturing the degree of foreign penetration in an industry (*fpind*) is measured by the share of foreign firms' employment in industry employment[9]. It can be gleaned from our previous argument in hypothesis 2 that this variable will not ordinarily impact positively on domestic firms' productivities unless it is adjusted for skill influence. Consequently it is interacted with the human capital factor to produce another variable termed *fpind*human k*. Unlike the expectation we have for the impact of its less refined counterpart this one should positively associate with productivity. Finally, to capture the influence of regional concentration of foreign firms we introduced the variable termed *agglo* which is defined as the proportion of sectoral employment accounted for by foreign firms in a region. Consistent with its motivation, if regional agglomeration matters for FDI Spillover, its coefficient should be positive.

Table 1: Summary Statistics

Variables	Observation	Mean	Std. Dev
output	455	2.66	1.79
output (dmstc)	302	3.01	2.56
k (capital)	455	2.97	1.68
l (labour)	455	3.34	0.33
m (material)	455	4.67	1.50
human k	455	2.82	1.97
invest in mach	455	0.49	0.67
export Intensity	455	-0.45	0.37
ecl	455	0.18	0.25
ecm	455	0.15	0.30
ecc	455	0.37	0.44
spillover (fpind)	455	1.43	1.14
spillover (fpp)	455	0.96	0.12
agglo	455	0.39	1.74

5. Results

Results of regressions estimated to address issues raised above are presented in Tables 2, 3 & 4. In the estimation that was carried out heteroscedasticity problem was dealt with by estimating variance-covariance matrices along the line suggested by White. From the output of all three regressions, the Ramsey RESET set assures us that misspecification should not be a major worry. Having said these, we can now proceed to comment on the results. Regression output for the own plant technology transfer (i.e. plant level channel) is first presented in Table 2 before estimates for possible industry wide spillover to domestic firms are presented in Table 3 (industry level channel). Table 4 reports the verdict for the regional effect. For all specifications the input variables are statistically significant and all turned up with expected signs.

Beginning with Table 2, results indicate that our variable of interest (*spillover (fpp)*) is positively signed and statistically significant at 1 per cent level. This suggests that domestic firms with foreign equity participation benefit from their cross border association. A ten-percentage point increase in foreign ownership increases firm productivity by 0.4 per cent. In other words, for an establishment where foreign participation increased from zero to ten per cent, productivity performance gains would be 0.4 per cent higher than for comparable domestic plants. We interpret this to mean a confirmation of our first hypothesis (hypothesis 1) which has also been affirmed in a previous study (Aitken and Harrison, 1999). The fact that sampled firms are reaping efficiency rewards from the presence of foreign assets in their equity mix is confirmative of the widely held, but controversially substantiated view that there is a technological spillover return to FDI[10].

Table 2: Foreign Direct Investment Spillover (Own-Plant Effect)

Variables	Plant Level
spillover(fpp)	0.04 (2.53)***
k (capital)	0.26 (2.58)***
l (labour)	0.11 (8.23)***
m (material)	0.19 (1.89)*
human k	0.12 (2.63)***
invest in mach	0.01 (1.74)*
export Intensity	-0.08 (1.80)*
ecm	0.96 (1.42)
ecl	-1.84 (1.58)
ecc	-1.01 (1.88)*
Adjusted R²	0.78
White Test	2.77***
Ramsey RESET	1.14
Observations	455

Notes: (1) The *t*-statistics are in parentheses; (2) Significant at ***1 per cent, **5 per cent and *10 per cent

Moving further to Table 3 we find that the coefficient of the spillover variable through the industry channel, though positive, is insignificant. This suggests that the positive impact of foreign capital on own plant productivity that was confirmed in hypothesis 1 above and would under normal circumstances be expected to diffuse to domestic firms effortlessly could not be confirmed. Failure to establish this sort of industry wide spread of foreign technology that some previous studies have established (Buckley *et al.*, 2007; Sinani and Meyer, 2004) is consistent in part, with the motivation of our hypothesis in 2. That is, in an environment where the requisite skills needed to facilitate the diffusion process is in short supply like South Africa, domestic enterprises are, to borrow Abramovitz's term, 'socially incapable' of internalising the externalised technological know-how of their foreign counterparts. It does appear that in general their inability to compete with foreign firms for the engagement of skilled employees have left them with the only option of relying on workforces that practically lack capacities to learn, copy, imitate and adapt the supposedly advanced practices of the foreign companies for their plants use.

To prove legitimacy of the above conclusion the argument put forward in hypothesis 2 further suggests that all domestic firms may not be equally unable to attract moderately skilled workforce that can help facilitate the technology appropriation process. As long as some of them are ready to make the right investment in human capital accumulation they should be able to increase their capacities to absorb and it follows that some amount of technology may diffuse to the locals through this group of firms. Thus if we can refine our analysis to take account of this fact we may observe a positive coefficient for the spillover variable. It turns out that after interacting the same spillover variable with human capital proxy the coefficient of the resulting variable (*spillover*human k*) is positive and significant at 5 per cent level suggesting that the diffusion process truly has a skill accent. Basically, what this finding confirms is that technology of resident MNCs does not spread effortlessly to domestic firms. A minimum level of skill capacity is required before successful absorption can take place. Thus our suspicion that skill matters for the spread of technology to domestic firms and that those with relatively more sophisticated workforce will be more successful in the absorption process is borne out in evidence

Finally, results from our exploration of the possibility of a regional dimension to the level of appropriated spillover by domestic firms as revealed in Table 4 below lend some faint support to our claim. The evidence is weakly present as estimated coefficient for the variable of interest (*agglo*) shows up with a positive value that is significant at barely 10 per cent.

Table 3: Foreign Direct Investment Spillover (Domestic Plant Effect)

Variables	Industry Level
spillover (fpind)	0.01 (1.53)
spillover (fpind)*human k	0.08 (1.96)**
k (capital)	0.02 (2.09)**
l (labour)	0.35 (8.03)***
m (material)	0.22 (2.11)**
human k	0.01 (1.09)
invest in mach	0.01 (1.38)
export Intensity	-0.27 (0.40)
ecm	1.74 (0.65)
ecl	1.03 (0.08)
ecc	-0.12 (1.81)*
Adjusted R²	0.72
White Test	2.19**
Ramsey RESET	0.11
Observations	302

Notes: (1.) The *t*-statistics are in parentheses; (2) Significant at ***1 per cent, **5 per cent and *10 per cent

Table 4: Foreign Direct Investment Spillover (Regional Effect)

Variables	Industry Level
Agglo	0.02 (1.71)*
k (capital)	0.09 (1.99)**
l (labour)	0.15 (4.00)***
m (material)	0.54 (4.51)***
human k	0.01 (0.74)
invest in mach	0.01 (1.00)
export Intensity	-0.25 (0.90)
ecm	1.12 (0.32)
ecl	0.9 (1.44)
ecc	-0.12(1.91)*
Adjusted R²	0.73
White Test	1.89**
Ramsey RESET	0.09
Observations	302

Notes: (1.) The *t*-statistics are in parentheses; (2) Significant at ***1 percent, **5 percent and *10 percent

6. Conclusions

Despite the nearly half a century old attempt to understand and explain the phenomenon of FDI spillover, the debate on whether or not host economies do benefit technologically from the activities of MNCs is still far from over. Precisely with respect to the role of country specific environmental factors that promote or inhibit FDI related technology transfer as well as the influence of regional factor evidence is especially lacking for South Africa. We have attempted in this paper to produce insight into this challenging problem by investigating the issues using firm level evidence.

Results from estimated regressions reveal that, despite the problem of skill supply deficit that the business environment imposes, firms with foreign equity participation are able to generate technological spillover for their host. Successful diffusion to domestic firms is however mediated and significantly circumscribed by the skill factor. Furthermore, evidence in support of the claim that level of FDI spillover relies on regional presence is present, though further research is needed to shed helpful insights that may be fed into policy learning.

From both analytical and policy perspectives these findings are instructive in the sense that they underscore the importance of unbundling spillover into its constituent channels of manifestation. At least in the context of South Africa, we now know that all firms are not equally inhibited by the environmentally imposed skill shortage constraint. Foreign firms are by virtue of their endowment able to get around it and deliver on their potentials. It is the domestic firms that are paying heavy technological price as a result. But whether this suggests a need for policy initiative to help domestic firms develop capacity to absorb spillover from MNCs should not be the issue. Rather what should concern us is finding how to increase the supply of human capital into the economy in general.

In the final analysis though, the message needs not be lost that the strength of our claim on the basis of revealed evidence is obviously weakened by limitations of the data we have used as well as econometric technique adopted. Robustness of cross-sectional based findings, we all know too well, is circumscribed by methodological inability to deal with such key challenges as endogeneity of factors, opportunistic investment behaviour, time variance etc. We cannot but admit being humbled by these considerations to say that evidence presented is at best suggestive and more sophisticated approaches especially the use of panel data would be required to take this further.

Notes

1   Some of these Reports include: The National Enterprise Survey (1998); The World Bank Large Manufacturing Firm Survey (1999); The World Bank Small, Medium and Micro Enterprise Firm Survey (1999); The Human Sciences Research Council (HSRC) Survey of Skills (2003). For a review, see Reza, C. D. (2007)

2   MacDougall (1960) is widely recognised as the first author to explicitly acknowledge technology spillover as one of the potential benefits from FDI in his analysis of the welfare effects of foreign investment. Other early authors who followed in the direction of McDougall to argue that productivity spillover is among the list of many positive impact of FDI were Corden (1967) and Caves (1971). In describing the works of these authors Blomstrom and Koko (1997, p. 10) noted: "The common aim of these studies was to identify the various costs and benefits of FDI and spillovers were discussed together with several other indirect effects that influence the welfare assessment, such as those arising from the impact of FDI on government revenue, tax policies, terms of trade, and the balance of payments. The fact that spillovers were taken into account was generally motivated by empirical evidence from case studies rather than by comprehensive theoretical arguments"

3   A number of authors have argued that MNCs pay higher wages than their domestic counterparts in order to attract and retain high quality staff (Lipsey and Sjoholm, 2004; Sinani and Meyer, 2004; Glass and Saggi 2002; Fosuri *et al.*, 2001; Aitken *et al.*, 1997).

4   This argument is only relevant if we are interested in vertical linkages which will involve a specification of inter-industry spillover model. In the present circumstance, our hypothesis and econometric specification are restricted only to intra-industry analysis.

5   According to one recent study economic activities in South Africa are geographically unequally distributed and concentrated with an estimated 70 percent of GDP being produced in only 20 per cent of places (Krugell *et al.*, 2005). Another work further finds that the phenomenon of industry concentration in certain provinces of the country is explained by provincial endowment in some aspects of economic attributes that are mentioned here. For instance, attributes such education, skill levels and road transport infrastructure density are shown to be important in the cases of Gauteng and KwaZulu-Natal while education and skill levels explains the case of the Western Cape Province (Naude, C. M. 2003).

6   See: https://www.enterprisesurveys.org/.

7   More that 80 per cent of respondents claim their establishments are single entities with no other separately owned operating establishments elsewhere in the country.

8   While it is tempting to argue that there is a prima-facie evidence, on the basis of this revelation, that foreign firms tend to be more productive than their domestic counterparts, the possibility that these firms (foreign) could just be engaging in opportunistic behaviours by investing only in the more productive sectors weakens the appeal of such argument.

9   Foreign firms are defined as firms with 10 per cent or more foreign equity participation.

10 This finding should however be interpreted with caution as decisions to invest in an economy by foreigners are sometimes driven by some perceived productivity potentials in a targeted establishment. It is therefore possible that the revealed performance superiority of these foreign owned enterprises is picking up this exogenous influence.

References

Aitken B. and Harrison A. (1999), 'Do Domestic Firms Benefit from Foreign Direct Investment: Evidence from Venezuela,' in *American Economic Review*, 89(3):605–618.

Barrios S. and Strobl E. (2002), 'Foreign Direct Investment and Productivity Spillovers: Evidence from the Spanish Experience,' in *Weltwirtschaftliches Archiv*, 138(3): 459–481.

Baskaran A. and Muchie M. (2009), 'Exploring the impact of National System of Innovation on the Outcomes of Foreign Direct Investment,' in *International Journal of Technological Learning, Innovation and Development (IJTLID)*, 2 (4): 314-345.

Blomstrom M. and Koko A. (1997), 'The Impact of Foreign Investment on Host Countries: A Review of the Empirical Evidence,' *World Bank Policy Research Working Paper*, No 1745. Available at: http://www.fetp.edu.vn/short-course/0203/Trade03/Readings/The%20impact%20of%20foreign%20investme nt%20on%20host%20countries.pdf (last accessed: 23 July 2009).

Blomstrom M., Kokko A. and Zejan M. (1994), 'Host Country Competition, Labour Skills, and Technology Transfer by Multinationals,' in *Weltwirtschaftliches Archiv*, 130(3): 521–533.

Buckley P. J., Wang C., and Clegg J. (2007), 'The impact of foreign ownership, local ownership and industry characteristics on spillover benefits from foreign direct investment in China,' in *International Business Review*, 16: 142–158.

Caves R. E. (1971), 'International Corporations: The Industrial Economics of Foreign Investment,' in *Economica*, 38 (3): 1-27.

Clarke G.R.G., Eifert B., Habyarimana J., Ingram M., Kapery W., Kaplan D., Schwartz M., and Ramachandran V. (2005), *South Africa: An Assessment of the Investment Climate*, Africa Private Sector Group, The World Bank, Available at: http://www.info.gov.za/otherdocs/2005/ dti_ica.pdf (last accessed 22 August 2009).

Corden W.M. (1967), 'Protection and Foreign Investment,' in *Economic Record*, 43: 209-232.

Crespo N. and Fontoura M. P. (2007), 'Determinant Factors of FDI Spillovers-What Do We Really Know?,' in *World Development*, 35(3): 410–425.

Das S. (1987), 'Externalities, and Technology Transfers through Multinational Corporations: A Theoretical Analysis,' in *Journal of International Economics*, 22: 171-182.

Fedderke J.W. and Romm A.T. (2006), 'Growth Impact and Determinants of Foreign Direct Investment into South Africa, 1956-2003,' in *Economic Modelling*, 23(5): 736-760.

Findlay R. (1978), 'Relative Backwardness, Direct Foreign Investment, and the Transfer of Technology: A Simple Dynamic Model,' in *Quarterly Journal of Economics*, 92(1): 1-16.

Fosuri A., Motta M. and Rønde T. (2001), 'Foreign Direct Investment and Spillovers through Workers' Mobility,' in *Journal of International Economics* 53 (1): 205–222.

Gelb S. and Black A. (2004), 'Globalisation in a Middle Income Economy: FDI, Production and the Labour Market in South Africa,' in William M. (ed.) (2004), *Labour and the Globalisation of Production*, London: Palgrave McMillan.

Girma S. (2003), 'Absorptive Capacity and Productivity Spillovers from FDI: A Threshold Regression Analysis,' *European Economy Group Working paper*, 25/2003, Available at: http://ideas.repec.org/a/bla/obuest/v67y2005i3p281-306.html (last accessed 23 July 2009).

Girma S. and Wakelin K. (2001), 'Regional Underdevelopment: Is FDI the Solution? A Semi-parametric Analysis,' *University of Nottingham GEP Research Paper* 2001/11, Available at: http://ideas.repec.org/p/cpr/ceprdp/2995.html (last accessed 23 July 2009).

Glass A. J. and Saggi K. (2002), 'Multinational Firms and Technology Transfer,' in *Scandinavian Journal of Economics*, 104 (4): 495–513.

Gorg H. and Greenway D. (2001), 'Foreign Direct Investment and Intra-Industry Spillovers: A Review of the Literature,' *The University of Nottingham Research Paper Series*, 2001/37, Available at: http://www.econstor.eu/bitstream/ 10419/2709/1/343806371.pdf (last accessed 23 July 2009).

Gorg H. and Strobl E. (2001), 'Multinational Companies and Productivity Spillovers: A Meta-Analysis,' in *Economic Journal*, 111(475): F723-F739.

Grant Thorton (2009), 'While the world bemoans a lack of orders, SA remains short of skills,' Available at: http://www.gt.co.za/News/Press-releases/International-business-report/2009/skills.asp (last accessed 23 July 2009).

Hermes N. and Lensink R. (2003), 'Foreign Direct Investment, Financial Development and Economic Growth,' in *Journal of Development Studies*, 40(1):142-163.

Kinoshita Y. (2001), 'R&D and Technology Spillovers through FDI: Innovation and Absorptive Capacity,' *CEPR Discussion Paper 2775*, Available at: http://ideas.repec.org/p/cpr/ceprdp/2775.html (last accessed 23 July 2009).

Koizumi T. and Kopecky K. J. (1977), 'Economic Growth, Capital Movements and the International Transfer of Technical Knowledge,' in *Journal of International Economics*, 7:45-65.

Kokko A. and Blomstrom M. (1995), 'Policies to Encourage Inflows of Technology through Foreign Multinationals,' in *World Development*, 23(3): 459–468.

Krugell W. F., Koekemoer G., and Allison J. (2005), 'Convergence or Divergence of South African Cities and Towns? Evidence from Kernel Density Estimates,' *Biennial Conference of the Economic Society of South Africa - "Development Perspectives: Is Africa Different?,"* Elangeni Holiday Inn, Durban, 7-9 September 2005, Available at: http://www.essa.org.za/download/2005Conference/Krugell.pdf (last accessed 25 August 2009).

Lipsey R. E. (2002), 'Home and Host Country Effects of FDI,' *NBER Working Paper 9293*, Available at: http://www.nber.org/papers/w9293.pdf (last accessed 23 July 2009).

Lipsey R. and Sjoholm F. (2004), 'Foreign Direct Investment, Education and Wages in Indonesian Manufacturing,' in *Journal of Development Economics*, 73(1): 415–422.

MacDougall G.D.A. (1960), 'The Benefits and Costs of Private Investment from Abroad: A Theoretical Approach,' in *Economic Record*, 36:13-35.

Marin A. and Bell M. (2004), 'Technology Spillovers from Foreign Direct Investment (FDI): an exploration of the active role of MNC subsidiaries in the case of Argentina in the 1990s,' *SPRU Electronic Working Paper Series 118*, Available at: http://www.sussex.ac.uk/spru/documents/sewp118.pdf (last accessed 9 August 2009).

Muchie M., and Baskaran A. (2009), 'The National Technology System Framework: Sanjaya Lall's Contribution to Appreciative Theory,' in *International Journal of Institutions and Economies*, 1 (1): 134-155.

Naude C. M. (2003), *Industry Concentration in South Africa*, Unpublished PhD Thesis, Available at: http://upetd.up.ac.za/thesis/available/etd-09092005-093957/unrestricted/00front.pdf (last accessed 25 August 2009).

Perez T. (1997), 'Multinational Enterprises and Technological Spillovers: An Evolutionary Model,' in *Journal of Evolutionary Economics*, 7(2): 169–192.

Ponomareva N. (2000), 'Are there Positive or Negative Spillovers from Foreign-owned to Domestic Firms?,' *New Economic School Working Paper*, BSP/00/042. Available at: http://www.nes.ru/english/research/pdf/2000/NPonomareva-e.pdf (last accessed 23 July 2009).

Renato G., Flores J. R., Maria P. F., and Rogerio G. S. (2007), 'Foreign Direct Investment Spillover in Portugal: Additional Lessons from a Country Study,' in *European Journal of Development Research*, 19(3):372–390.

Reza C. D. (2007), 'Skills Shortages in South Africa: A Literature Review,' *Development Policy Research Unit DPRU Working Paper*, 07/121, Available at: http://www.commerce.uct.ac.za/research_units/dpru/Work ingPapers/PDF_Files/WP_07-121.pdf (last accessed 23 July 2009).

Sinani E. and Meyer K. E. (2004), 'Spillovers of technology transfer from FDI: the case of Estonia,' in *Journal of Comparative Economics*, 32(3): 445-466.

Torlak E. (2004), 'Foreign Direct Investment, Technology Transfer, and Productivity Growth in Transition Countries: Empirical Evidence from Panel Data,' *Cege Discussion Paper* 26. Available at: http://wwwuser.gwdg.de/~lstohr/cege/Diskussionspapiere/26_Torlak.pdf (last accessed 23 July 2009).

Wang J. and Blomstrom M. (1992), 'Foreign Investment and Technology Transfer: A Simple Model,' in *European Economic Review*, 36:137–155.

World Bank (2004), *South Africa: An Assessment of the Investment Climate*, Available at: http://www.info.gov.za/otherdocs/2005/dti_ica.pdf (last accessed 23 July 2009).

Xu B. (2000) 'Multinational enterprises, technology diffusion, and host country productivity growth,' in *Journal of Development Economics*, 62(2): 477–493.

Yudaeva K., Kozlov K., Malentieva N., and N. Ponomareva (2003), 'Does foreign ownership matter? The Russian Experience,' in *Economics of Transition*, 11(3): 383–409.

African Journal of Science, Technology, Innovation and Development
Vol. 1, No. 1, 2009
pp. 103-135

Innovations Systems in Renewable Natural Resource Management and Sustainable Agriculture: a literature review[1]

Laxmi Prasad Pant* and Helen Hambly-Odame**

Abstract

The innovation systems approach to renewable natural resource management and sustainable agriculture has been recognized as a new conceptual tool. However, there is a lack of systematic review of literature in this emerging field of inquiry. This review of over 100 relevant contributions concludes that there have been two major paradigm shifts in agricultural science, technology, innovation and development – from the transfer of technology approach to the various research and extension systems, and then to innovation systems. Innovation systems in agriculture have two major theoretical foundations – innovation systems in manufacturing, and various systems approaches in agriculture. The review revealed that it is imperative to establish a new generation of innovation systems research in renewable natural resources that integrates three apparently distinct fields of studies – environmental management, international development and innovation studies.

Key Words: agriculture; natural resources; innovation systems; international development; paradigm shift; low-income countries.
JEL Classification: Q10, Q16, Q18

* Researcher, School of Environmental Design and Rural Development, University of Guelph, Guelph, Ontario N1G 2W1, Canada. Email: lpant@uoguelph.ca

** Associate Professor, School of Environmental Design and Rural Development, University of Guelph, Guelph, Ontario N1G 2W1, Canada. Email: hhambly@uoguelph.ca

1. Introduction

Agricultural research and rural development has moved from technology transfer thinking to various systems approaches with varying degrees of integration of local and indigenous, and expert ways of learning and innovation, the latest concept being innovation systems. Scholarly discussion on the innovation systems thinking has been influenced by two bodies of literature: one in manufacturing under the work of Edquist (1997; 1999), Freeman (1987) and Lundvall (1992a; 2002), and the other in rural development and agriculture, such as the work of Biggs *et al.* (2003; 2004), Clark *et al.* (2001; 2003), Hall and Nahdy (1999), Hall (2001), Hall *et al.* (2001a; 2000; 2001b; 2003b), Sumberg (2005) and Temel *et al.* (2003). Recently the innovation system framework has gained substantial acceptance in international development discourse among a diverse group of stakeholders, including multi-lateral and bi-lateral donors, actors of national innovation systems, and local and indigenous communities.

Despite encouraging advancements in agricultural science, technology and innovation, the world has witnessed food crises time and again. The major causes of the 21st century food crisis are identified as climatic change and extreme weather events, environmental pollution and natural resources degradation, soaring oil prices and diversion of food grains to biofuels, and rising food demand, trade restrictions and commodity market speculations (Katz, 2008; von Braun, 2007). The impacts of the food crisis are most severe on vulnerable sections of society in both developed and developing countries that have a low level of adaptation capacity - local and indigenous as well as expert knowledge-based adaptation. Although the causes and consequences of the food shortages are increasingly apparent, solutions are contentious around the paradox of productivity and sustainability. This paper argues that neither productivity-focused interventions nor sustainability-focused interventions in the past have been helpful to generate social welfare, specifically poverty reduction and social inclusion in low-income countries, and thus it is imperative to revamp the conventional discourse on international development.

This paper first characterizes the systems approaches to innovation for development, and then discusses two major paradigm shifts in renewable natural resources and agriculture, first from the technology transfer to research and innovation systems and second towards innovation system thinking. Finally, research gaps are identified in the emerging literature on innovation systems in renewable natural resource sector, and specifically smallholder agriculture.

2. Paradigm Shifts in Science, Technology, Innovation and Development

Before embarking on the discussion of processes of innovation, it is important to understand how paradigm shift takes place in science, technology, innovation and rural development. Here "paradigm" is defined as a shared belief system or worldview. Thomas Kuhn (1962) differentiates between the transition through a paradigm and the transformation of a paradigm or paradigm shift. Transition through a paradigm involves adjustments to conceptual tools within a paradigm. But when such adjustments are not enough to deal with contemporary, often intractable problems such as food crisis, an entirely new paradigm with a new set of tools is necessary. Specifically in agriculture, we find two major paradigm shifts: (1) from the technology transfer to research and extension systems; (2) from the research and extension systems to innovation systems (Table 1).

Table 1: Key paradigm shifts in agriculture

Attributes	Technology transfer	Research & extension systems	Innovation systems
1. Approaches			
Science-society relationship	Science and technology are relatively independent of political and other social factors.	Research takes place in a historically defined social, political, economic, agro-climatic context.	Innovation takes place in a historically defined social, political, economic, agro-climatic context.
Trajectory of technological change	Specific, unambiguous (e.g., researcher, an innovator, extensionist, a carrier and farmer, an adopter).	Not an unambiguous, one-way progression. Multiple actors as a part of research and extension systems.	Not an unambiguous, one-way progression. Multiple actors as a part of innovation systems.
Treatment of time	Classification of a technology as modern and traditional (e.g., modern and traditional crop varieties).	A technology is made up of many old and new components.	A technology and innovation are made up of many old and new, often unknown, components.
2. Contexts			
Organizational structure	National Research Institutes, National Extension Institutes.	National Agriculture Research Systems, including extension and education systems.	National Agriculture Research Systems, including extension and education systems, and private sector organizations.
Institutional responsibility	Research and extension as the public sector business.	Need for public and private partnerships in research and extension.	Need for public and private partnerships for sector coordination.
Role of public sector extension	Technology dissemination. The release/control of information.	Systems management. Active participants in applied research.	Systems management. Potentially a central node among the wide range of nodes in an innovation system.
3. Social goals			
Policy environment	Rigid state of equilibrium between public and private sector; regulative policy instruments.	Flexible state of disequilibrium among various institutions, state's role as facilitator of research systems.	Flexible state of disequilibrium among various institutions, state's role as facilitator of innovation systems.
Role of a farmer	The end user of a technology.	An experimenter within research systems.	An expert innovator within innovation systems.

Source: Authors with reference to Biggs (1990), Thrupp and Altieri (2001), Sulaiman and Hall (2002) and World Bank (2007).

2.1 Transformation from linear to systems approaches

The technology transfer approach emphasizes linear relationships between science and society. Examples of this paradigm were commodity programs, such as crop specific research, technology development and extension advocated during the era of the *Green Revolution* (1960s and 70s). Focus was on the development of modern varieties of few major cereals, access to irrigation and use of chemical inputs. Standardization of technology as a package of practices was the major effort of agricultural research and extension overlooking temporal and spatial variations in farming communities, their local and indigenous practices, and diversity of agro-ecological conditions. The diffusion of technological innovation was presumed to be a linear process: researchers working in modern scientific establishments as the producer of knowledge and information, the extension worker as carriers of information and the farmer as the ultimate adopter (Rogers, 2003).

Since the linear paradigm considers science and technology relatively independent of historical, social, political, cultural and other institutional factors, various systems approaches have emerged to integrate various types and sources of knowledge and innovations, including local and indigenous knowledge and practices (Biggs, 1990; Sulaiman and Hall, 2002; Thrupp and Altieri, 2001; World Bank, 2007). In addition to the historical, social, cultural and political factors, attempts have also been made towards integrating social and biophysical systems because innovation in renewable natural resources and agriculture depends on the interdependence of three major sub-systems – technological systems, social systems and biophysical systems. Addressing such interdependence would result in three major competencies:
(1) favourable agroecological conditions; (2) access to agricultural technology, such as a crop variety; and (3) a range of support services – credit, inputs, produce markets, processing facilities, insurance against risk of crop failure, and favourable institutional environments for life-long learning and innovation. However, learning-based competencies of stakeholders are replacing the conventional sources of core competencies – agroecological comparative advantage and technological advancements – providing a rationale for a paradigm shift towards innovation systems approaches (Pant *et al.*, 2008).

In response to the increasing acceptance of systems approaches, standalone education, research and extension institutes are increasingly required to operate as a part of a broader systems emphasizing broad-based public-private collaboration, involving formal sector actors and rural communities in the processes of learning and innovation. The role of public agricultural extension as an authority-in-charge of release and control of technology and information is also being revised as a systems manager and facilitator of public-private collaboration, and community-based research and development interventions (Sulaiman and Hall, 2002; Sulaiman *et al.*, 2005).

Various systems approaches appeared to guide rural and agricultural development, particularly in the context of low-income countries. We identify three major approaches: (1) systems approaches to technological innovations; (2) systems approaches to the diffusion of innovation; and (3) systems approaches to individual livelihood[2].

(a) Systems approaches to technological innovations in agriculture

Recognizing the limitation of the linear paradigm of technological innovations, Farming System Research Extension (FSR&E) was proposed during the 1960s and early 1970s (Collinson, 2000). The major attributes of the farming systems approach, as it was initiated during the time, are engagement of an interdisciplinary team of practitioners, involvement of small farmers, appreciation of their knowledge and practices and use of small farms as the main unit of analysis. The work on FSR&E is charac-terized by systematic description of system goals, system boundaries, linkages, feedback loops, goal seeking mechanisms, diagnosis of problems, design of experiments, and evaluation of results (Collinson, 1987; Zandstra *et al.*, 1981). The evaluation of experiments can result in three outcomes – either to reformulate on-station experiments to plan further cycle of on-farm research or to disseminate results through extension agencies (Collinson and Lightfoot, 2000).

Farming systems thinking has substantially evolved since its begin-ning in the 1960s. Two major paths of evolution are evident from the literature. First, it evolved in terms of system scale, performance criteria and target beneficiaries (Hart, 2000). In terms of scale, it changed from the original focus on cropping systems to crop livestock integrated systems in the 1980s and then at community and watershed levels in 1990s. In other words, FSR&E either has a predetermined focus on selected commodities, a whole farm focus, a natural resource systems focus including on-farm and off-farm activities, or a livelihood systems focus which includes on-farm, off-farm and non-farm activities (Norman, 2002). Similarly the original focus of *productivity* as a system performance indicator had been modified to include stability in the 1980s and then further included *sustainability* in the 1990s. A sustainable system should be economically viable, socially and politically acceptable, environmentally benign, technologically adaptable, and institutionally manageable (Farshad and Zinck, 2001; Filson and Duke, 2004; Filson *et al.*, 2003). The original focus on small farmers as target beneficiaries has been changed to address social inclusion, such as gender, cultural consideration and generational equity.

Second, the farming systems thinking evolved in terms of its focus on social processes of technological innovation. The first generation of FSR&E involved research into farming systems, the second as systems research into farming, and the third as critical farming systems (Bawden, 2002). As Bawden (2002) mentions, philosophical jargons equivalent to the three generations are respectively the hard systems appropriate for exploring strategies for technological change, soft systems appropriate for exploring human activities and critical systems aiming to address the power asymmetry and equity[3].

The latest farming systems thinking urges integrating hard systems, soft systems and critical systems (Bawden, 2002). In generic terms, the focus on linkages and recycling between biophysical components at farming systems level, such as using crop residues as livestock feeds and composting crop fields by livestock manure, has evolved into linkages beyond biophysical components to include socio-economic parameters at various levels. However, integrating biophysical and socio-economic variables is complicated because the effective ways to investigate how some of these variables interact at different spatial and temporal scales are still emerging (Goss *et al.*, 2004).

(b) Systems approaches to diffusion of innovation

The diffusion of agricultural innovation has long been perceived as a linear process: research as knowledge producers, extension as knowledge disseminators, and farmers as the passive recipients of the information being disseminated (Rogers, 2003). While FSR&E attempts to conceptualize technological innovation in systems perspectives, Agricultural Knowledge and Information System (AKIS) attempts to conceptualize diffusion of innovation from systems perspective. Information consists of a pattern imposed on data which simultaneously influences the interpretation of those data and enables them to be transmitted, while knowledge consists of a meaningful experience (Röling, 1990). Knowledge cannot be transmitted to others unless encoded or transformed into information. Knowledge generated in one component of the system is transformed into information for use in the other component of the system. Niels Röling (1990; 1994) proposed the AKIS and emphasized institutional pluralism for knowledge production, diffusion and application. His definition of the AKIS is as follows:

> AKIS is a set of agricultural organizations and/or persons, and the links and interactions between them, engaged in such processes as the generation, transformation, transmission, storage, retrieval, integration, diffusion and utilization of knowledge and information, with the purpose of working synergistically to support decision making, problem solving and innovation in a given country's agriculture or a domain thereof (Röling, 1990: 1).

While there are still apparent challenges to work with multiple stakeholders reconciling expert knowledge systems and local practices (Ammann, 2007; Pant and Hambly- Odame, forthcoming), the AKIS thinking lately embraces a holistic, cross-sectoral and multi-occupational nature of rural livelihoods with a revised focus on rural development and environmental management (FAO, 2000; Rivera, 2004; Röling and Jiggins, 1998). Furthermore, with the changing relationships between public and private actors, the revolution in information and communication technologies (ICTs), and new concepts of interactive learning and innovation provide new challenges to production, exchange and application to knowledge pertaining to agriculture and related activities (FAO, 2000; Hambly-Odame, 2003).

(c) Systems approaches to individual livelihoods

The small-farm focus of FSR&E thinking was reiterated by the holistic and people-centred approach of sustainable livelihood thinking because renewable natural resource-based activities are one of the important sources of livelihoods for small farmers in low-income countries (Chambers and Conway, 1992). The sustainable livelihood systems focus on assets, capabilities and activities required for a means of living (Brock, 1999; Cromwell, 1999; Niehof, 2004; Scoones, 1998)[4]. Sustainable livelihoods are, in fact, the emergent properties of soft systems as it recognizes how assets endowments are converted into capital entitlements (Bawden and Packam, 1989; Blaikie *et al.*, 2002; Sen, 1999)[5]. The ability to convert asset endowments to resource and capital entitlements is influenced by a host of micro- and macro-economic contexts including policy, agreements, and conventions (Murray, 2001).

Depending on their ability to foresee vulnerability in the immediate future, three types of livelihood strategies are available for smallholder farmers in low-income countries (Brock, 1999). First, farmers either diversify or simplify agriculture depending on the agro-ecological comparative advantage. Second, if their livelihoods remain insecure from natural resource-based activities, both on-farm and off-farm, farmers expand into non-farming activities while staying on their own farm to increase income. Third, if they still fail to secure livelihoods, the last resort would be migration, which may be seasonal or even permanent. In a nutshell, farmers would generate livelihood outcomes, such as food security and well-being, through balanced use of on-farm, off-farm and non-farm livelihood strategies (Norman, 2002; Norman *et al.*, 1995). Rural livelihood systems are increasingly characterized as cross-sectoral, multi-occupational and multi-locational, and thus the new transition is from conventional focus on individual livelihoods towards community-based livelihood interventions (Ellis and Biggs, 2001).

In summary, although there has been a transition within each of the above systems approaches, the set of conceptual tools provided by each paradigm have been inadequate in addressing key problems in low-income countries – economic downturn, climate change, environmental pollution, poverty, hunger and social exclusion – that transcend disciplinary, sectoral, institutional and hierarchical mandates. It would be appropriate to transform these systems approaches to developing an entirely new set of conceptual tools to address the increasing complexity in agriculture, renewable resource management and rural development. The new source of core competence lies on our ability to learn and innovate in response to changing technological, economic, social, climatic, and environmental challenges and opportunities.

2.2 Transformation towards the innovation systems approach

Innovation systems in agriculture evolved within two distinct scholarly traditions: (1) innovation in manufacturing sector; and (2) innovation in renewable natural resource sector.

(a) *The Industrial Revolution and subsequent emergence of innovation system thinking*

Although innovation systems thinking is relatively new, the concept of innovation is centuries old. Francis Bacon's essay on 'The New Atlantis', published posthumously in 1626, is argued to be the very first text to examine the processes of innovation (Wallance, 1982). The early Industrial Revolution in Western Europe was foreseen by Bacon almost a century earlier (Bacon, 1901)[6]. It was also speculated upon by Adam Smith in 'The Wealth of Nations' (1776), of course, under the influence of Bacon (Smith, 1978). However, the mid-20th century theorists were still preoccupied by the technology transfer paradigm largely due to the massive economic, political, and socio-cultural changes brought about by the Industrial Revolution and its precedents of capitalist expansion and imperialism in Latin America, Asia and Africa (Wallerstein, 1979). One of the first attempts to investigate the changes brought about by the Industrial Revolution from systems perspective was the work of the Science and Technology Policy Research (SPRU) at The Freeman Centre, University of Sussex, U.K. (e.g. Freeman, 1987; Patel and Pavitt, 1997, Pavitt, 1999).

The innovation systems thinking is also linked to Staffan Linder, a liberal economist and a former conservative minister in the Swedish Government (Lundvall *et al.*, 2002). Linder (1961)'s essay on 'Trade and Transformation' inspired Bengt-Åke Lundvall, and his colleagues at Aalborg University, Denmark to work on innovation systems thinking. A definition of innovation system is as follows:

> The narrow definition would include organizations and institutions involved in searching and exploring – such as R&D departments, technological institutes and universities. The broad definition...includes all parts and aspects of the economic structure and the institutional set-up affecting learning as well as searching and exploring – the production system, the marketing system and the system of finance present themselves as sub-systems in which learning takes place (Lundvall, 1992b: 12).

The World Bank (2007:xiv) defines an innovation system "...as a network of organizations, enterprises and individuals focused on bringing new products, new processes, and new forms of organization into social and economic use, together with the institutions and policies that affect their behaviour and performance." In essence, an innovation system is the network of public and private actors engaged in production, exchange, regulation and application of knowledge in a particular domain of economic activity. The application of knowledge and technology to create new products, new processes, and new organizational forms requires a range of support services, such as credit, inputs, insurance schemes and social safety nets that are well beyond the scope of research systems. Thus nation states are responsible for providing favourable public policy environments for advancements of science, technology and innovation for development.

Lundvall *et al.* (2002) assert that as long as nation states exist as political entities with their own agendas to innovate, it is useful to work with the concept of national systems. This notion is analogous to that which Friedrich List (1841) argued about national systems of production in his best known work 'The National System of Political Economy.' Lundvall *et al.* (2002) further argues that regionalization and globalization of the economy definitely weakens national capacity to innovate, especially when the knowledge exchanges are tacit and difficult to codify, such as local and indigenous ways of learning and innovation. Stakeholders who originate from different cultures, such as scientists trained through formal schooling and local and indigenous communities, may have a hard time engaging in interactive learning and innovation (Lundvall *et al.*, 2002). Hence, there may be tension between pan-regional and global integration of expert knowledge systems through patenting and licensing, and appreciation of diversity in local and indigenous knowledge and practices at the national and sub-regional levels (Sumberg, 2005).

Finally, the arguments for qualifying innovation systems thinking turn out to be pragmatic in the realm of public policy. On the one hand, although nation states are responsible for national public policies, this does not preclude the importance of regional and global policies. Indeed, the process of regionalization and globalization of innovation makes it even more pertinent to understand the role of nation states (Freeman, 1987; Lundvall, 1992b). Therefore, nation states are the basic unit of analysis for addressing complexity of regional and global systems. On the other hand, nation states vary at national as well as sub-regional levels. Focus on national, pan-regional and international levels overlook the socio-economic impacts of interventions at the local level. However, given the proximity of firms or farms in a local cluster or virtual networks, innovation would occur whether they compete or complement each other (Burger *et al.*, 2001).

(b) Sectoral transformation towards the innovation systems in agriculture

After the advent the 21st century, the humanity not only inherited the challenges from the previous centuries, such as negative environmental, social and political impacts of the Industrial Revolution but also new challenges emerged and opportunities opened due to global changes in climate, environment, technology, politics, economies and increasingly greater interdependence of nation states. One approach to address the challenges and opportunities in low-income countries has been the innovation systems approach in renewable natural resources and agriculture because it is realized that symmetric economic growth in countries where people are overwhelmingly dependent on renewable natural resource-based livelihoods would come through investment in agriculture (World Bank, 2008; IAASTD, 2008).

An application of the innovation systems thinking in agriculture implies transition through the innovation systems framework from one sector to another. However, the research and extension systems in agriculture requires a transformation – a "paradigm shift" – because the set of tools provided by these systems, including various transitions through the farming systems, knowledge and information systems, and livelihood systems discussed above, have been inadequate to address the challenges of the 21st century. Nevertheless in the context of the origin of the innovation systems concept, four major limitations are apparent in applying the innovation systems framework in agriculture in low-income countries. Firstly, the focus should be on systems building in addition to its application on systems analysis because in most developing countries innovation systems do not exist (Lundvall *et al.*, 2002). Despite consider-able investment this limitation is still apparent in agriculture, because the initial efforts were focused on diagnostic studies to demonstrate t h e relevance of innovation systems approach in agriculture (Rajalahti *et al.*, 2008; Ekboir and Hambly-Odame, 2007). Secondly, systems building processes should systemically recognize the innovative capacity and epistemological backgrounds of all relevant stakeholders, including socially excluded women, youth, small-scale producers and resource-poor farmers (Shiva, 1997; Stamp, 1989). Such concerns specifically relate to the question of power and the influence of power structure on various ways of learning and innovation (Lundvall *et al.*, 2002).

Thirdly, it is important to develop processes to facilitate stakeholder interaction involving the public and non-profit private sectors, on the one hand, and the for-profit private and informal sectors, on the other, because the innovation systems framework has evolved from manufac-turing where actors are predominantly from the corporate sector (Edquist, 1997; Lundvall, 1992a; Nelson, 1993). Thus public-private partnerships should also involve informal sector stakeholders, including local and tribal communities. Finally, unlike the case of manufacturing, agricultural production cycles are longer and influenced by technological, biophysical, cultural, political and socio-economic factors. There is an apparent challenge to address the generic differences between firms and farms, specifically when agriculture is primarily practised for subsistence in resource constrained contexts (Clark *et al.*, 2003).

In short, conceptually rural development and agriculture traversed two major paradigm shifts – first, from the technology transfer to the research and extension systems, and then to innovation systems. These paradigm shifts brought about parallel changes in organizational and institutional arrangements, and provide new challenges to managing economic, social and environmental changes in global agriculture.

3. Innovation Processes in Agriculture

The first paradigm shift from technology transfer to research and extension systems has reflected systemic reforms in policy and practices, such as a move from the training and visit (T&V) systems of extension to more inclusive programs such as on-farm trials and farmers field schools (FFSs) integrating expert knowledge-based learning and experimentation with local knowledge and practices. However, these reforms were more of *science and technology (S&T) policy* changes in agriculture. The second paradigm shift towards the innovation systems calls for reforms in *innovation policy*, including changes in *S&T* policy.

3.1 Innovation in research and extension systems

In the past, the technology transfer paradigm influenced policies enabling delivery of technological messages to rural farming communities. Research, extension and educational organizations were established to facilitate agricultural research and extension, often with a focus on commodity programmes for major crops, such as research and extension in rice, wheat and maize. Government agencies were committed to facilitating farmers' access to land, farm machinery, high yielding crop varieties, fertilizers, pesticides, irrigation and credit. In South Asia, for example, the T&V system was promoted until the 1990s (Sulaiman and Hall, 2002). The T&V system institutionalized regular monthly workshops involving research scientists and extension officers. The extension officers in turn conducted fortnightly training for village level extension workers. From regular trainings and occasional farm visits, extension officers used to bring back first-hand experience of rural farming situations. This linear approach assumes that fundamental research is the source of innovation, and formally organized public extension systems and mass media pass it to farming communities, often ignoring local and indigenous knowledge and practices (Mundy and Compton, 1995).

When negative impacts of the *Green Revolution* technologies were apparent, practitioners and theorists questioned the linear paradigm of agricultural research and extension, leading towards the emergence of various systems approaches with various degrees of integration of expert knowledge-based interventions, and local and indigenous knowledge and practices. Under the influence of the FSR&E approach, researchers introduced the practice of on-farm trials. Researchers employ various levels of on-farm trials based on the stakeholder involvement at planning and implementation stages (Norman *et al.*, 1995). First, trials are moved to farmers' fields, but these are planned and implemented by researchers, and known as *researcher planned and implemented trials*. Usually there are rental contracts between researchers and farmers to use designated pieces of land.

The second type of on-farm trial is *researcher planned and farmer implemented*. In this case, farmers are the passive recipients of the research-ers' ideas to be tested. They still participate as hired labours. A third type of trials is *farmer planned and implemented*, where farmers have a greater role in decision making. As we move along this continuum of the typology of trials, there is increasing participation of farmers in decision making and relatively greater opportunity to integrate expert and local ways of learning and experimentation. However, most of the on-farm trials are limited to farmers' participation for material incentive as they participate by providing resources, such as land and labour in return for food, cash and other material incentives (Pretty, 1993). The objectives of on-farm trials are to demonstrate the benefits of modern technology, such as yield potential of modern crop varieties over traditional varieties, and persuade farmers to adopt modern technologies. In reality, however, local and indigenous technologies, such as local crop varieties, often perform better than modern ones, particularly under farmers' manage-ment practices, and in absence of external inputs, such as chemical fertilizers and pesticides (Biggs, 1983). Recognizing this flaw of expert knowledge-based learning and experimentation, local crop varieties and associated knowledge were integrated in the formal research systems as early as the mid-1990s (Fujisaka, 1995). However, small changes without a transformation across the research and extension systems had very limited development impact.

A revised focus on farmers' participation in agricultural research and extension has been achieved through farmers' field schools, a kind of "discovery learning" through collaboration of extension workers and farmers in key areas such as pest management, nutrient management, water management and crop improvement. Farmers' participation in such processes has become relatively more functional because they participate by forming groups with their specific interest in discovery learning and problem solving (Pretty, 1993). Contrary to individual learning in the technology transfer approach, systems approaches encourage social learning through interaction among stakeholders, integrating expert knowledge-based learning and innovation with local and indigenous ways of learning and innovation (Leeuwis, 2004). As far as social learning is concerned, crop improvement, for example, typically involves four stages – selection of source germplasm, trait development, cultivar development and varietal testing (Morris and Bellon, 2004). A complete participatory plant breeding (PPB) involves farmers at all stages while efficient PPB involves farmers at germplasm selection and varietal evaluation. At the extremes of the continuum, researchers either work independently from farmers at all stages as was common in the transfer of technology approaches or farmers save their own seeds, often through mass selection based on physical appearance, and local knowledge and perception, irrespective of the scientific research and development (R&D) (Ceccarelli *et al.*, 2000; Humphries *et al.*, 2005; Morris and Bellon, 2004).

3.2 Innovation in innovation systems

A range of case studies that employed innovation systems thinking revealed that farmers' participation in the research process is necessary but not sufficient for production, exchange and application of knowledge held among myriad of stakeholders (Table 2). Review of the case studies clearly resulted in two broad categories of the innovation systems study in agriculture: (1) technology-specific studies; and (2) commodity-specific studies. The first category of inquiry begins with the development of *process technology*, such as agricultural biotechnology and PPB. Processes technology oriented studies focused on farmers' participation in research and technology development often with an intermediary role of non-governmental organizations (NGOs) (Clark *et al.*, 2002; Joshi *et al.* 2005). The role of the for-profit private sector was relatively informal. In Andhra Pradesh India, conventional biotechnology, such as tissue culture, bio-fertilizer and bio-pesticides, were promoted in the short-term while the long-term strategy had been on molecular biotechnology and genetic engineering. In the Chitwan district of Nepal, participatory plant breeding (PPB) had been promoted and efforts were made to reduce the dichotomy between participatory and centralized plant breeding practices, often subjecting PPB varieties and farmers varieties into centralized varietal testing procedures, such as on station disease screening and multi-location testing (Pant and Hambly-Odame, forthcoming). In this effort, PPB has been redefined as client-oriented breeding depending on the involvement of local and indigenous communities (Witcombe *et al.*, 2006; Witcombe *et al.*, 2005).

Table 2: Case studies on innovation in innovation systems

Process technology
1. Agricultural biotechnology led by public university in India under the support of the Dutch Ministry of Foreign Affairs (Clark et al., 2002).
2. Participatory rice improvement led by civil society organizations in Nepal under the support of the UK Department for International Development (Joshi et al. 2005).

Commodity production for domestic uses
1. Small-scale food processing led by civil society organizations in Bangladesh without external support (World Bank, 2007).
2. Tomato production and post-harvest led by civil society organizations in India under the support of the UK Department for International Development (Clark et al., 2003).
3. Potato production and post-harvest led by the government agencies and civil society organizations in Peru under the support of the Swiss Agency for Development and Cooperation (SDC) (Devaux et al., 2007).

Commodity production for export markets
1. Vanilla production and post-harvest led by farmers' cooperatives in India without external support (World Bank, 2007).
2. Pineapple production and post-harvest led by the government agencies in Ghana under the support of the German Agency for Technical Cooperation , the U.S. Agency for International Development and the Natural Resources Institute, UK (World Bank, 2007).
3. Cut flowers production and post-harvest led by a farmers' association in Columbia without external support (World Bank, 2007).
4. Floriculture led by the government agencies and civil society organizations in Kenya under the support of international research organizations, such as International Centre of Insect Physiology and Ecology (ICIPE) and CABI (Bolo, 2007).

Commodity production for both domestic and export markets
1. Mango production and post-harvest let by the government agencies in India under the support of the UK Department for International Development (Hall et al., 2003a).
2. Medicinal plant and Ayurvedic medicine led by civil society organizations and private companies in India without external support (World Bank, 2007).
3. Cassava production and processing led by government agencies and civil society organizations in Colombia without external support (World Bank, 2007).
4. Cassava production and processing led by government agencies in Ghana without external support (World Bank, 2007).
5. Shrimp production and post-harvest led by civil society organizations in Bangladesh without external support (World Bank, 2007).

The second and most common category of study of innovation systems in agriculture begins with a focus on *commodity production and post-harvest processing*, either for domestic uses, export markets or both. Production and post-harvest management of tomato in India and potato in Peru illustrate how NGOs and farmers' organizations facilitate innovation systems when small-holder crop producer in niche specific agro-ecological conditions target domestic markets. Tomatoes from the foothills of the Himalayas (Clark *et al.*, 2003), and potatoes from the Andes (Devaux *et al.*, 2007) carry brand names that are recognized in the domestic markets in respective countries. The diversity of Peruvian potato cultivars facilitates pro-poor innovation systems to function because diversity does not work with monoculture and agricultural mechanization preferred by large-scale producers. In these cases, NGOs facilitate sector coordination often with a challenging task of involving stakeholders from the for-profit private sector.

Crop production for export markets has more challenges than production for domestic market or household consumption. Anticipating new produce demands under changing consumer preference in a distant market, increasingly stringent regulations, and price competition from other countries are a few typical challenges. For example, Indian vanilla could not compete with artificial vanilla in the international market; Ghanaian pineapple growers switched to new varieties and involved themselves in value-added products including cut and sliced fruits and certification for fair trade and organic produce; Colombian cut flower producers shifted from carnations and chrysanthemums to spray carnations and roses, and from stems to ready-made bouquets (World Bank, 2007); and Kenyan flower exporters experimented with sea shipment protocols and market diversification of cut-flowers to increase price competitiveness (Bolo, 2007).

In India, interaction between major actors, such as vanilla farmers, public research and support agencies and private companies was missing. In Ghana, the success of farmers' owned export companies established by the government was limited although the government supports non-traditional exports, such as through a free trade zone establishing certification mechanisms with specific training on pesticide management. In both Colombia and Kenya, the success of the floriculture sector was limited by physical and emotional distance between the flower sector and research community leading to weaker interaction of expert knowledge-based interventions, and local and indigenous practices leading to over reliance on international R&D for which farmers have to pay costly fees and royalties.

Targeting production and processing for the domestic as well as the export market creates an additional challenge of bringing small-scale and large-scale entrepreneurs together. Crop production and post-harvest for domestic uses has to compete with the large-scale production and processing for export as lower quality produce that are not suitable for export remains in the domestic market. For example the small-scale food processing in Bangladesh experienced competition from large-scale food processors which affected the livelihoods of poor food processors, mainly women (World Bank, 2007). The specific challenges of small-scale food processors include: (1) insufficient technical skills to meet hygiene standards and consumer preferences; (2) credit markets and tax incentives deliberately designed for large-scale processors, and (3) the physical and emotional isolation from urban centres. Specifically the food sector coordination was hindered by physical and social isolation of small-scale food processors from urban consumers, service providers, and large-scale processors in spite of supportive interventions from non-governmental organizations (NGOs).

In the Indian mango sector, an unsuccessful attempt was made to address the divides between small-scale producers and large-scale producers through the establishment of farmers' cooperative federation and their collaboration with other stakeholders for sea freight of mangoes to London, U.K. (Hall *et al.*, 2003a). Initially perceived to be a technical problem, the unacceptable quality of mangoes arriving in London was realized, only after three years, as the missing links between pre-harvest and post-harvest service agencies to effectively manage anthracnose, a quality-related disease. In a similar case, a small number of large pharmaceutical companies in India developed associated research, design, and marketing organizations in herbal medicine with strong links to the international market to increase their market share (World Bank, 2007). However local health practitioners and rural communities holding the ethno-botanical knowledge and practices under the philosophy of Ayurveda were marginalized. Not only did the Medicinal Plant Board of India failed to promote effective sector coordination but also non-profit private sector actors, such as NGOs and private foundations, were unsuccessful in coordinating traditional health care workers and rural development workers towards integrating expert knowledge-based herbal medicine, and local and indigenous practices of health care. The sector coordination was hindered by a long-established mistrust among public research organizations, the private sector and NGOs, coupled with philosophical differences between scientific and traditional medicine, and the resistance to integration and interaction of multiple stakeholders.

Cassava processing in Colombia and Ghana, to produce animal feed and starch, also provided lessons on how one should deal with the divide between small-scale and large-scale producers. The relative success of the sector in Colombia relied on stakeholders' willingness to explore different forms of partnerships and long-established traditions of cooperative movement and industry associations (World Bank, 2007). However, the focus on research *per se* was rarely extending linkages beyond the research system. Likewise, in Ghana, the Ayensu Starch Company Limited (ASCo) provided infrastructure and incentives to the private sector, facilitated interaction between actors, and coordinated and supported a network of small-scale cassava producers (World Bank, 2007). The interaction was, however, classically top-down and therefore could not facilitate effective knowledge sharing.

In addition to the duality between small-scale and large-scale producers, the Bangladeshi shrimp industry was required to comply with increasingly stringent markets, such as the US Food and Drug Administration's Hazard Analysis Critical Control Point (HACCP) sea food regulation, address environmental protection and social equity considerations associated with shrimp production (World Bank, 2007). NGOs provided technical support to poor shrimp farmers but they were inefficient in linking shrimp farmers with the research systems, markets and domestic policy process.

The above case studies revealed that success of an agricultural intervention lies on actor structure facilitating the interaction of resources (natural, financial, human, social and human made capitals), knowledge management processes including integration of expert knowledge systems, and local and indigenous knowledge and practices, and stakeholder values around economic growth, social inclusion and environmental protection (Christensen *et al.*, 2004, Pant, forthcoming). In other words, a key challenge for learning and innovation was fostering broad-based sector coordination beyond the conventional boundaries of the public, non-profit private, for-profit private and informal sectors. One of such challenges is bridging the socio-psychological divide between scientific research, and local and indigenous practices (Pant and Hambly-Odame, forthcoming). Value systems of stakeholders, individual as well as collective, are influenced by how public and private stakeholders behave within an organization and across the innovation system

(Durham, 1991). A deviation from the long-established value systems and organizational culture pushing the conventional boundaries, of course for positive changes across innovation systems, would provide leverage to sector coordination (Casson, 2003; Pant and Hambly-Odame, forthcoming; Sternin and Choo, 2000; Williams, 2002).

4. Towards a New Generation of Innovation Systems Research

In the retrospect, the mid-20th century *Green Revolution* in developing countries brought a many-fold increase in productivity of rice, wheat and maize. This success resulted through international collaboration and government commitments to farmers' access to land, farm machinery, high yielding crop varieties, fertilizers, pesticides, irrigation and credit (Conway, 1997; Parayil, 1992). However, in general poor farmers in arid and marginal areas who were already vulnerable hardly realized any benefits, notably the African continent was largely bypassed. Moreover, the *Green Revolution* caused some negative environmental impacts, such as soil degradation, air and water pollution, and loss of biodiversity and associated knowledge and practices. Biodiversity loss is a kind of memory loss because local and indigenous knowledge get lost with it (Nazarea, 1998). The negative social and environmental impacts were the unintended consequences of well-meaning interventions to feed starving people in low-income countries, for example, the Bengal Famine in India (1942-1943) with four million deaths in 1943 alone (Axinn, 1988; Singh, 1999). In response to these negative social and environmental repercussions, sustainable agricultural practices, such as multiple cropping, biodiversity conservation and low-external input agriculture, were promoted as early as 1970s under the rubric of FSR&E (Pretty, 1993).

In addition to the mid-20th century challenges that are still being unsuccessfully addressed under the paradigm of research and extension systems, new sets of *actors* and *factors* are emerging in the 21st century, posing seemingly intractable challenges to achieve global prosperity. Private corporations (PCs) and NGOs are becoming increasingly active often challenging the *status quo* of the public sector. For example, PCs in life sciences are at the forefront of modern agricultural science and technology, specifically biotechnology and genetic engineering, often referred to as the *Gene Revolution* (Parayil, 2003). Biotechnology and genetic engineering have a huge potential to breed stress tolerant, disease resistant, and nutrient fortified crop varieties, and disease tolerant animal breeds, but, today's investments in biotechnology, concentrated in the private sector and driven by commercial interests, have had limited impacts on smallholder productivity, specifically in low-income countries because the research systems are further detached from the local and indigenous knowledge and practices (World Bank, 2008). Activist NGOs are influencing government to enact stringent legislations in agriculture to advance positive social and environmental changes, but stringent environmental legislations are not free from unintended consequence, such as a reduction in agricultural productivity.

Along with socio-economic factors, climatic and environmental *factors* are rapidly changing. Climate change, for example, has increasingly affected, and has been affected by, agriculture as a result of a *positive feedback* cycle of carbon emitting industrial agricultural practices further amplifying climate change (IPCC, 2007). Sustainable agricultural practices, such as agro-forestry, home gardens, perennial crops and minimum tillage, are either carbon sequestering or carbon neutral, thus stabilizing greenhouse gas emission through a *negative feedback cycle.* Moreover, environmental pollution and natural resource degradation as a result of unsustainable agricultural practices not only jeopardizes human health and ecological integrity putting more pressures on healthcare systems, but also reduces agricultural productivity, thus limiting the supply of safe and adequate foods (IAASTD, 2008). Thus one of the intractable challenges of the 21st century is to increase agricultural productivity while ensuring ecological integrity and human well-being. One way to address this challenge is to explore complementarities between *science and technology policy,* and *environmental policy* (OECD, 2005). Specifically in international development, *social policy* is equally important. However, the motivations for these policy portfolios are different: *S&T policy* is traditionally driven by productivity and national competitiveness; *environmental policy* is driven by international agreements and concerns for our planet; and *social policy* is motivated by welfare or equity goals, such as providing social safety nets to vulnerable sections of society until their adaptation capacity is sufficiently developed. However, such an adaptation capacity cannot be developed through replacement of local and indigenous adaptation practices but through an integration of expert knowledge-based adaptation strategies with that of local and indigenous communities (Bhattacharyya, 2006).

Global change literature, specifically in climate and environmental change, has moved beyond its conventional focus on assessing change and determining ecological and human vulnerability, and begun a scholarship of developing local *adaptation capacity*, a convergence with the literature in rural and international development (Adger, 2006; Gallopin,

2006; McLaughlin and Dietz, 2008; Smit and Wandel, 2006). However, specifically in renewable natural resource management and agriculture, the conventional tools provided by the research and extension paradigms of innovation, including various transition throughout the last decades, have consistently failed to address the challenges and opportunities created by rapidly changing actor and factor structures.

Since the tools provided by research and extension systems were inadequate to address the challenges of the 21st century in low-income countries, it is imperative to transform the research and extension paradigms into innovation systems paradigm, moving beyond the conventional focus on research systems, and also facilitating learning and innovation to respond to rapidly changing contexts (World Bank, 2007; Rajalahti *et al.*, 2008; Rajalahti, 2009). The new generation of research on innovation systems, therefore, should move beyond technology-specific as well as commodity-based interventions as are evident from the case studies on innovation in innovation systems. In addition to these two types of interventions, a third category of potential interventions would focus on consumer concerns, such as carbon neutral and/or carbon sequestering agriculture, low-external inputs agriculture, advancements of smallholder agriculture, sustainable use of forest products and medicinal herbs, and fair trade practices (Carlsson *et al.*, 2002). These interventions are neither within a domain of a particular disciplinary research and extension systems nor under a mandate of a commodity programme, and thus the contemporary challenge is to develop a new generation of innovation systems research that investigates the convergence of three, seemingly distinct, fields of interdisciplinary studies: *environmental management, international development* and *innovation studies*.

One way forward to achieve the convergence is to focus on innovation capacity development in renewable natural resource management and sustainable agriculture while addressing the seemingly intractable social problems, such as food crisis, hunger and social exclusion. Building the *capacity to adapt*, also referred to as *capacity to innovate* within a broader framework of innovation systems, entails the development of collective context-specific skills, practices, routines, institutions, and policies to put existing and new knowledge, including local and indigenous knowledge, into productive use in response to changing technological, economic, social, climatic, and environmental challenges and opportunities (Hall, 2005). The capacity to innovate is an emergent property of a system that comes through the interrelationships and interactions among various elements of the system (Morgan, 2005). In practice, the thinking on capacity development and innovation systems intersects on two critical points – fostering sector coordination, and facilitating interactive learning and innovation (Hambly-Odame *et al.*, 2007; Pant and Hambly-Odame, 2006).

5. Conclusions

Clearly, there have been two major paradigm shifts in agricultural science, technology, innovation and development. The first paradigm shift was from technology transfer to various research and extension systems, and the second was from research and extension systems to innovation systems. Although systems research had been introduced in agriculture as early as the mid-20th century, public sector predominance in research and extension is still evident in many low-income countries. New ways of managing rural prosperity have been increasingly sought in response to the development of new technologies, such as transgenic agriculture, agro-energy production, and ICTs, and emergence of new actors including non-profit private organizations, private foundations and private companies. The new ways of managing rural prosperity involve not only the scientific research and technological innovations, but also how stakeholders learn in response to seemingly unknown and uncertain economic, social, climatic and environmental changes.

The available case studies on the analysis of the innovation systems in agriculture demonstrated that the collaboration of stakeholders with apparently conflicting value systems is the greatest hindrance to building innovation systems. In other words, the greatest challenge in managing global prosperity involves collaboration among stakeholders with seemingly conflicting goals, such as enhancing economic growth, maximizing social welfare and equity, and minimizing environmental damage. When stakeholders are required to work at the interface of environmental management, international development, and innovation studies, the challenge to address the paradox of productivity and sustainability would be formidable unless the issue of social equity at the bottom of the pyramid is addressed. An innovation systems approach that integrates management, development and innovation is critical for better policy outcomes and learning. Therefore, we need the innovation systems mode of framing agriculture and natural resource interventions.

Notes

[1]   The authors would like to acknowledge the International Development Research Centre (IDRC) of Canada for funding this work.

[2]   The authors appreciate that a fair treatment of these well-established systems approaches demands a book length, but the aim in this paper is to provide a context to the evolution of the innovation systems in agriculture.

[3]   Theorists characterize systems thinking in terms of the processes of inquiry: hard systems recognize the world as systemic while soft systems view the processes of inquiry as systematic (Bawden, 2002; Checkland, 2000).

[4]   *Asset* is that which one has while *resource* is that which one uses (Niehof, 2004). When resources are used to create new resources, they are called *capital*.

[5]   Emergent properties are the properties visible only at the systems levels.

[6]   The changes during the *Industrial Revolution* speculated by Bacon and reiterated by Smith included disciplinary education, the establishment of specialized laboratories, and improvements in techniques of measurement and experimentation to increase the cognitive efficiency of discovery, invention and innovation.

References

Adger W. N. (2006), 'Vulnerability,' in *Global Environmental Change,* 16:268-281.

Ammann K. (2007), 'Reconciling Traditional Knowledge with Modern Agriculture: A Guide for Building Bridges,' in Krattiger A., Mahoney R. T., Nelsen L., Thomson J. A., Bennett A. B., Satyanaryana K., Graff G. D., Fernandez C. and Kowalski S. P. (eds.), *Intellectual Property Management in Health and Agricultural Innovation: A Handbook of Best Practices,* Oxford, U.K.: Davis; U.S.A.: MIHR/PIPRA.

Axinn G. H. (1988), 'International Technical Interventions in Agriculture and Rural Development: Some Basic Trends, Issues, and Questions,' in *Agriculture and Human Values,* 5:6-15.

Bacon F. (1901), *The New Atlantis* (Written in 1626), New York: P. F. Collier & Son.

Bawden R. (2002), 'On the Systems Dimension in FSR,' in *Journal of Farming Systems Research and Extension,* 5:1-18.

Bawden R. J., and R. Packam (eds.) (1989), *Improving Agriculture through Systemic Action Research,* Richmond (NSW): University of Western Sydney.

Bhattacharya A. (2006), 'Using participatory GIS to bridge knowledge divides among the Onge of Little Andaman Island, India,' in *Knowledge Management for Development,* 2:97-110.

Biggs S. D. (1983), 'Monitoring and control in agricultural research systems,' in *Research Policy*, 12(1):37-59.

_____(1990), 'A Multiple Source of Innovation Model of Agricultural Research and Technology Promotion,' in *World Development* 18:1481-1499.

_____(2003), 'An Actor Innovation Systems Approach to Scaling-up of Agricultural Technologies,', pp. 35-54, in Proceeding of a Workshop Uptake Pathways and Scaling-up of Agricultural Technologies to Enhance the Livelihoods of Nepalese Farmers, Kathmandu, Nepal: MOAC/NARC/NARDF/DFID-HARP.

Biggs S. D. and Harriet M. (2004), 'Strengthening Poverty Reduction Programmes Using an Actor-oriented Approach: Examples from Natural Resources Innovation Systems,' *AgREN Network Paper*, No.134.

Blaikie P., Cameron J., and Sedona D. (2002), 'Understanding 20 Years of Change in West-Central Nepal: Continuity and Change in Lives and Ideas,' in *World Development* 30:1255-1270.

Bolo M. (2007), 'Applying Innovation Systems to Agricultural Science, Technology and Innovation: The case of Kenya's floriculture industry,' in *An Expert Workshop on Enhancing Agricultural Innovation Systems*, March 22-23, 2007, Washington, D. C.: The World Bank.

Brock K. (1999), 'Implementing a Sustainable Livelihoods Framework for Policy-Directed Research: Reflections form Practice in Mali,' *IDS Working Paper*, No. 90., Institute of Development Studies, Brighton.

Burger K., Daniel K., and Henry S. (2001), 'Clustering of Small Agro-Processing Firms in Indonesia,' in *International Food and Agribusiness Management Review*, 2:289-299.

Casson, M.C. (2003). *The Entrepreneur: An Economic Theory*, Second Edition, Northampton, MA, USA: Edward Elgar.

Carlsson B., Jacobsson S., Holmen M., and Rickne A. (2002), 'Innovation systems: analytical and methodological issues,' in *Research Policy* 31:233-245.

Ceccarelli S., Grando S., Tutwiler R., Baha J., Martini A.M., Salahieh H., Goodchild A., and Michael M. (2000), 'A methodological study on participatory barley breeding I. Selection Phase,' in *Euphytica*, 111:91-104.

Chambers R. and Conway G. (1992), 'Sustainable rural livelihoods: practical concepts for the 21st Century,' *IDS Discussion Paper*, No. 296.

Checkland P. (2000), 'Soft Systems Methodology: A Thirty Year Retrospective,' in *Systems Research and Behavioural Sciences*, 17(S1):S11-S58.

Christensen C. M., Scott D. A., and Erik A. R. (2004), *Seeing What's Next: Using the Theories of Innovation to Predict Industry Change*, Boston, Massachusetts: Harvard Business School Press.

Clark N. (2001), 'Innovation system, Institutional Change and the New Knowledge Market: Implications for Third world Agriculture Development,' *UNU/INTECH Discussion Paper*, No. 10, Maastricht, The Netherlands.

Clark N., Hall A., Sulaiman R. V., and Naik G. (2003), 'Research as Capacity Building: The Case of an NGO Facilitated Post-Harvest Innovation System for the Himalayan Hills,' in *World Development*, 31(11):1845-1863.

Clark N., Yoganand B., and Hall A. (2002), 'New science, capacity development and institutional change: the case of the Andhra Pradesh-Netherlands Biotechnology Programme (APNLBP),' in *The International Journal of Technology Management and Sustainable Development*, 1(3):196-212.

Collinson M. E. (1987), 'Farming Systems research: Procedures for technology development,' in *Experimental Agriculture*, 23: 365-386.

_____(2000), 'Introduction,' in Collinson M. E. (ed.), *A History of Farming Systems Research*, Oxon, UK/ Rome, Italy: CABI Publishing/FAO.

Collinson M. E. and Lightfoot C. (2000), 'The Future of Farming Systems Research,' in Collinson M. E. (ed.), *A History of Farming Systems Research*, Oxon, UK/ Rome, Italy: CABI Publishing/FAO.

Conway G. (1997), The Doubly Green Revolution: Food for all in the 21st Century, New York: Comstock Publishing Associates.

Cromwell E. (1999), 'Agriculture, biodiversity and livelihoods: issues and entry points,' in *Final Report*, London: Overseas Development Institute.

Devaux A., Ordinola M., Manrique K., Lopez G., and Thiele G. (2007), 'Stimulating pro-poor innovation within market chain of native potatoes: A case of Peru,' in *An Expert Workshop on Enhancing Agricultural Innovation Systems*, March 22-23, Washington, D. C.

Durham W. H. (1991), *Coevolution: genes, culture, and human diversity*, Stanford, California: Stanford University Press.

Edquist C. (ed.) (1997), Systems of Innovation: Technologies, Institutions, and Organizations, London; Washington: Pinter.

Edquist C. and Hommen L. (1999), 'Systems of Innovation: theory and policy for the demand side,' in *Technology in Society*, 21:63-79.

Ekboir J. and Hambly-Odame, H. (2007), *A Learning Alliance in Agricultural Innovation Systems: Report of a Scoping Mission*. Washington D.C.: Agricultural and Rural Development Division, The World Bank.

Ellis F., and Biggs S. D. (2001), 'Evolving Themes in Rural Development 1950s-2000s,' in *Development Policy Review*, 19:437-448.

FAO (2000), Agricultural Knowledge and Information Systems for Rural Development (AKIS/RD): Strategic Vision and Guiding Principle, Rome: FAO/The World Bank.

Farshad A. and Zinck J. A. (2001), 'Assessing Agricultural Sustainability Using the Six-Pillar Model: Iran as a Case Study,' in edited by Gliessman S. R (ed.), *Agroecosystem Sustainability: Developing Practical Strategies*, New York: CRC Press.

Filson G. C. and Duke C. (2004), 'Integrating Farming Systems Analysis of Intensive Farming,' in Filson G. C. (ed.), *Intensive Agriculture and Sustainability: A Farming Systems Analysis*, Vancouver: UBC Press.

Filson G. C., Pfeiffer W. C., Paine C., and Taylor J. R. (2003), 'The Relationship Between Grand River Dairy Farmers' Quality of Life and Economic, Social and Environmental Aspects of Their Farming Systems,' in *Journal of Sustainable Agriculture*, 22:61-77.

Freeman C. (1987), Technology Policy and Economic Performance: Lessons from Japan, London: Pinter.

Fujisaka S. (1995),'Incorporating Farmers' Knowledge in International Rice Research,' in Warren D. M., Slikkerveer L. J., and Brokensha D. (eds.), *The cultural Dimension of Development: Indigenous knowledge systems*, London: Intermediate Technology Publications.

Gallopin G. C. (2006), 'Linkages between vulnerability, resilience and adaptive capacity,' in *Global Environmental Change*, 16:293-303.

Goss M. J., John R. O., Filson G. C., Barry D. A., and Olmos S. (2004), 'Developing Predictive and Summative Indicators to Model Farming Systems Components,' in Filson G. C, *Intensive Agriculture and Sustainability: A Farming Systems Analysis,* Vancouver: UBC Press.

Hall A. (2005), 'Capacity Development for Agricultural Biotechnology in Developing Countries: An Innovation Systems View of What it is and How to Develop it,' in *International Journal of Development*, 17(5): 611-630.

Hall A. (2001), 'The Development and Use of the Innovation Systems Framework in India: Northern Perspective,' in *Paper presented in the conference North-South Research Co-operation*, Royal Netherlands Academy of Arts and Science, The Netherlands.

Hall A., Bockett G., Taylor S., and Sivamohan M. V. K. (2001a), 'Why Research Partnership Really Matter: Innovation Theory, Institutional Arrangements and Implications for Developing New Technology for the Poor,' in *World Development* 29(5): 783-797.

Hall A., Clark N., Taylor S., and Sulaiman R. V. (2001b), 'Institutional learning through technical projects: Horticultural technology R&D systems In India,' *AgREN Network Paper*, No. 111.

Hall A., Clark N., Sulaiman R. V., Sivamohan M. V. K., and Yogandnd B.(2000), 'New Agendas for Agricultural Research in Developing Countries: Policy Analysis and Institutional Implications,' in *Knowledge, Technology, and Policy*, 13: 70-91.

Hall A., and Nahdy S. (1999), 'New methods and old institutions: The systems context of farmer participatory research in national agricultural systems. The case of Uganda,' in *AgREN Network Paper*, No. 93.

Hall A., Sulaiman R. V., Yoganand B., Raina R. S., Clark N., and Naik G. (2003a), 'Institutional learning and change: towards a capacity-building agenda for research. A review of recent research on post harvest innovation systems in South Asia,' in Hall A., Yoganand B., Sulaiman R. V., and Clark N. (eds.), *Post-harvest innovations in innovation: reflection on partnership and learning*, Patancheru, India: International Crops Research Institute for the Semi-Arid Tropics (ICRISAT).

Hall A., Sulaiman R. V., Clark N., and Yoganand B. (2003b), 'From measuring impact to learning institutional lessons: an innovation systems perspective on improving the management of international agricultural research,' in *Agricultural Systems*, 78: 213-241.

Hambly-Odame H. (2003), 'Connecting the Two Stations of Agricultural Research and Rural Radio.' *Journal of Development Communications*. 23(1): 1-14.

Hambly-Odame H., Fitzsimons J., Pant L. P, and Sykanda P. (2007), 'Innovations in Farm Families and Rural Communities: Capacity Development for Broadband Use in Southern Ontario,' *Working Paper*, Spring, Ontario Ministry of Government Services/University of Guelph, Toronto/Guelph.

Hart R. (2000), 'FSR - Understanding Farming Systems,' in Collinson M. E. (ed.), *A History of Farming Systems Research*, Oxon, UK/ Rome, Italy: CABI Publishing/FAO.

Humphries S., Gallardo O., Jimenez J., Sierra F., and members of the Association of CIALs of Yorito Sulaco and Victoria (2005), 'Linking small farmers to the formal research sector: lessons from a participatory bean breeding programmed in Honduras,' *AgREN Network Paper*, No. 142.

IAASTD (2008), International Assessment of Agricultural Science and Technology for Development, Washington, D.C.: Island Press.

IPCC (2007), 'Climate Change 2007: Synthesis Report,' in *Fourth Assessment Report of the Intergovernmental Panel on Climate Change (IPCC)*, Cambridge: Cambridge University Press.

Joshi, K. D., D. Biggs S. D., Gauchan D., Devkota K. P., Devkota C. K, Shrestha P. K., and Sthapit B. R. (2005), 'The evolution and spread of socially responsible technical and institutional innovations in a rice improvement system in Nepal,' *Discussion Paper 8*, CAZS Natural Resources, University of Wales, Bangor, Wales, UK.

Katz S. (2008), 'Food to fuel and the world food crisis,' in *Anthropology Today*, 24:1-3.

Kuhn T. S. (1962), *The Structure of Scientific Revolutions*, Chicago: The University of Chicago Press.

Leeuwis C. (2004), Communication for Rural Innovation: Rethinking agricultural extension, Third Edition, Oxford: Blackwell Science.

Linder, S. B. (1961), *An Essay on Trade and Transformation*, Stockholm: Almqvist & Wicksell.

List F. (1841). *The National System of Political Economy*, Reprinted 1966, New York: A.M. Kelley.

Lundvall B.-Å. (1992a), National Systems of Innovation: towards a theory of innovation and interactive learning, London: Pinter Publishers.

Lundvall B.-Å. (1992b), 'Introduction,' in Lundvall B.-Å. (ed.), *National Systems of Innovation: Towards a Theory of Innovation and Interactive Learning*, London: Pinter Publishers.

Lundvall B.-Å., Johnson R., Andersen E. S., and Dalum B. (2002), 'National systems of production, innovation and competence building,' *Research Policy*, 31: 213-231.

McLaughlin P. and Dietz T. (2008), 'Structure, agency and environment: Toward an integrated perspective on vulnerability,' in *Global Environmental Change*, 18:99-111.

Morris M. L., and Bellon M. R. (2004), 'Participatory plant breeding research: Opportunities and challenges for the international crop improvement,' in *Euphytica*, 136: 21-35.

Morgan P. (2005), The Idea and Practice of Systems Thinking and Their Relevance for Capacity Development, Maastricht: European Centre for Development Policy Management.

Mundy P. A., and Compton J. L. (1995), 'Indigenous Communication and Indigenous Knowledge,' in Warren D. M., Slikkerveer L. J., and Brokensha D. (eds.), *The cultural Dimension of Development: Indigenous knowledge systems*, London: Intermediate Technology Publications.

Murray C. (2001), 'Livelihoods research: some conceptual and methodological issues,' *Chronic Poverty Research Centre-Background Paper*, No. 5, University of Manchester, UK.

Nelson R. R. (1993), *National Innovation Systems: A Comparative Study*. Oxford: Oxford University Press.

Niehof A. (2004), 'The significance of diversification for rural livelihood systems,' in *Food Policy*, 29 (4): 321-338.

Norman D. W. (2002), 'The Farming Systems Approach: A Historical Perspective,' *Paper Presented at the 17th IFSA Symposium*, Lake Buena Vista, Florida.

Norman D., Worman F. D., Siebert J. D., and Modiakgotia E.. (1995), 'The Farming Systems Approach to Development and Appropriate Technology Generation,' in *FAO Farming Systems Management Series 10*, Rome: FAO.

OECD (2005), *Governance of Innovation Systems - Synthesis Report,* Volume 1, Paris: Organization for Economic Co-operation and Development (OECD).

Pant L. P. and Hambly-Odame H. (2006), 'Multi-stakeholder deliberation on dialectical divides: an operational principle of the Systems of Innovation,' in *Knowledge Management for Development Journal,* 2:60-74.

Pant L. P., Hambly-Odame H, Hall A., and Sulaiman R. V. (2008), 'Learning Networks Matter: Challenges to Developing Learning-Based Competence in Mango Production and Post-Harvest in Andhra Pradesh, India,' *UNU-MERIT Working Papers,* No. 2008-069.

Pant L. P. (forthcoming), 'Learning networks for bridging knowledge divides in international development: approaches and initiatives,' *IKM Working Paper,* European Association of Development Research and Training Institutes (EADI), Bonn, Germany.

Pant L. P. and Hambly-Odame H. (forthcoming), 'The promise of positive deviants: bridging divides between scientific research and local practices in smallholder agriculture,' in *Knowledge Management for Development Journal.*

Patel P. and Pavitt K. (1997), 'The Technological Competencies of the World's Largest Firms: Complex and Path-Dependent, but not much Variety,' in *Research Policy,* 26 (2): 141-156.

Pavitt K. (1999), *Technology, Management and Systems of Innovation.* Cheltenham, U.K.: Edward Elgar.

Parayil G. (1992), 'The Green Revolution in India: A Case Study of Technological Change,' in *Technology and Culture,* 33(4): 737-756.

_____(2003), 'Mapping technological trajectories of the Green Revolution and the Gene Revolution from modernization to globalization,' in *Research Policy,* 32 (6): 971-990.

Pretty J. N. (1993), 'Alternative Systems of Inquiry for a Sustainable Agriculture,' *IDS Bulletin,* 25 (2): 37-48.

Rajalahti R., Janssen W., and Pehu E. (2008), 'Agricultural Innovation Systems: From Diagnostics toward Operational Practices,' in *Agriculture and Rural Development Discussion Paper 38,* Washington, D. C.: The World Bank.

Rajalahti R. (2009), 'Promoting Agricultural Innovation Systems Approach: The Way Forward,' Available at: http://knowledge. cta.int/en/content/view/full/9564 (Accessed on 24 August 2009).

Rivera W. (2004), 'Communication for Rural Development: Challenge to Diffuse Development Information on Non-agricultural Rural Needs,' *Paper presented in the Ninth United Nations Roundtable on Communication for Development,* 6-9 September, FAO, Rome.

Rogers E. M. (2003), *Diffusion of Innovation,* Fifth Edition, New York: Free Press.

Röling N. G. (1990), 'The Agricultural Research-Technology Transfer Interface: A Knowledge Systems Perspective,' in Kaimowitz D. (ed.), *Making the Link: Agricultural Research and Technology Transfer in Developing Countries*, London: Westview Press.

_____(1994), 'Agricultural Knowledge and Information Systems,' in Blackburn D. J. (ed.), *Extension Handbook: process and practices*, Toronto: Thompson Educational Publications.

Röling N. G. and Jiggins J. (1998), 'The ecological knowledge system,' in Röling N. G. and Wagemakers M. (ed.), *Facilitating Sustainable Agriculture: Participatory Learning and Adaptive Management in Times of Environmental Uncertainty*, Cambridge, UK: Cambridge University Press.

Scoones I. (1998), 'Sustainable rural livelihoods: a framework for analysis,' *IDS Working Paper*, No. 72.

Sen A. (1999), *Development as Freedom*, Oxford: Oxford University Press.

Shiva, V. (1997), *Biopiracy: the plunder of nature and knowledge*, Toronto: Between The Lines.

Singh D. (1999), 'The green revolution and the evolution of agricultural education and research in India,' in *Genome*, 42: 557-561.

Smit B. and Wandel J. (2006), 'Adaptation, adaptive capacity and vulnerability,' in *Global Environmental Change*, 16(3): 282-292.

Smith A. (1978), *The Wealth of Nations*, New York: J. M. Dent & Sons Ltd.

Stamp P. (1989), Technology, Gender, and Power in Africa, Ottawa: IDRC.

Sternin J. and Choo R. (2000), 'The Power of Positive Deviancy,' in *Harvard Business Review*, 78 (1): 14-15.

Sulaiman V.R., and A. Hall. (2002). "Beyond technology dissemination: reinventing agricultural extension." *Outlook on Agriculture* . 31:225-233.

Sulaiman V.R., Hall A., and Bharathi N. S. (2005), 'Effectiveness of Private Sector Extension In India and Lesson for the New Extension Policy Agenda,' *AgREN Network Paper*, No. 141.

Sumberg J. (2005), 'Systems of Innovation theory and the changing architecture of agricultural research in Africa,' in *Food Policy*, 30 (1): 21-41.

Temel T., Janssen W., and Karimov F. (2003), 'Systems analysis by graph theoretical techniques: assessment of the agricultural innovation system of Azerbaijan,' in *Agricultural Systems*, 77(2): 91-116.

Thrupp L. A. and Altieri M. (2001), 'Innovative Models of Technology Generation and Transfer: Lessons Learned from the South,' in Wolf S. and Zilberman D. (ed.), *Knowledge Generation and Technological Change: Institutional Innovation in Agriculture*, Boston: Kluwer Academic Publishers.

von Braun J. (2007), 'The World Food Situation: New driving Forces and Required Actions,' in *Food Policy Report*, Washington, D.C.: International Food Policy Research Institute (IFPRI).

Wallance A. F. C. (1982), The Social Context of Innovation. Bureaucrats, Families, and Heroes in the Early Industrial Revolution, as Foreseen in Bacon's New Atlantis, New Jersey: Princeton University Press.

Wallerstein I. (1979), *The Capitalist World-economy*, London: Cambridge University Press.

Williams P. (2002), 'The Competent Boundary Spanner,' in *Public Administration*, 80(1): 103-124.

Witcombe J. R., Gyawali S., Sunwar S., Sthapit B. R. and Joshi K. D. (2006), 'Participatory Plant Breeding is Better Described as Highly Client-oriented Plant Breeding. II. Optional farmer Collaboration in the Segregating Generations,' in *Experimental Agriculture*, 42(1): 79-90.

Witcombe J. R., Joshi K. D., Gyawali S., Musa A. M., Johansen C., Virk D. S., and Sthapit B. (2005), 'Participatory Plant Breeding is Better Described as Highly Client-oriented Plant Breeding. I. Four Indicators of Client - orientation in Plant Breeding,' in *Experimental Agriculture*, 41(3): 299-319.

World Bank (2007), Enhancing Agricultural Innovation: How to Go Beyond the Strengthening of Research Systems, Washington, D. C.: The World Bank.

_____(2008), World Development Report 2008: Agriculture for Development, Washington, D.C.: World Bank.

Zandstra H. G., Price E. C., Litsinger J. A., and Morris R. A. (1981), *A Methodology for On-farm Cropping Systems Research*, Los Banos: IRRI.

African Journal of Science, Technology, Innovation and Development
Vol. 1, No. 1, 2009
pp. 136-166

Are North-South Technological Spillovers Substantial? A dynamic panel data model estimation

Watu Wamae*

Abstract

This paper argues that *actual* technological spillovers are not substantial in developing countries because of the weak absorptive capacities. A panel data analysis is used in an attempt to gain insight into the specific aspects that enable economies to benefit from the backlog of existing knowledge. The findings indicate that low productivity effects of human capital coupled with weak or virtually non-existent systems of innovation are at the root of the observed ambiguity with regard to the spillover gains that are expected to play a significant role in sparking growth.

Keywords: absorptive capacity, spillovers, developing countries, systems of innovation.
JEL Classification: C33, O24, O47, O57

1. Introduction

Amongst developing countries, there is a growing rift between the few economies that have managed to "take-off" and the overwhelming majority that is increasingly being marginalised by the current economic trend of rapid transformations. From a more general perspective, there is a great deal of evidence against the inevitable convergence predicted by earlier models, such as Solow (1956). Temple (1999:151) points out that, "Poor countries are not catching up with the rich, and to some extent the international income distribution is becoming polarized." This situation has arisen with technology taking the centre stage in driving economies

---

* Development, Policy and Practice, The Open University, Walton Hall, Milton Keynes, MK7 6AA. Email: watu.wamae@gmail.com

and modifying dynamics in the global economy. The question addressed in this paper is: What lies behind the ability of a handful of developing countries to catch up with industrialised countries while the vast majority recedes further into marginalisation?

Technology-led growth is characterised by rapid changes, due to pressure from such factors as rapid technical change and liberalisation, and evidence suggests that the returns to human capital are increasing, resulting in skill-biased technical change. However, the primary focus of classical, neoclassical and endogenous growth theory remains the allocation of scarce resources; structural feedback mechanisms that determine the dynamism of linkages and synergies in a rapidly changing environment are not taken into account. The national systems of innovation is an alternative approach proposed within the evolutionary technical change framework.

Pioneered and elaborated by Nelson and Winter (1982), Rosenberg (1983), Freeman (1987) among others, the national systems of innovation approach emphasises that the innovation process is a process of interactive learning in which actors improve their competences.[1] The endogenous structural, institutional and social factors, which constitute the so-called technological gap, have been stressed within the systems of innovation approach as largely responsible for driving economies apart. The underlying fact is that rapid economic transformations render competence acquisition increasingly tacit, and hence the importance of an adequate system of networks and linkages between and amongst actors and institutions.

This paper attempts to show how the wide divergence amongst economies is mirrored by the rate of growth of knowledge, and that it reflects structural, institutional and social factors. More specifically, it is argued that domestic innovation in developing countries is a vital source of sustainable growth despite the popular view that importing high technology equipment is the best way or even the only way to ignite growth in developing countries, and especially in the poorest, since they hardly invest in domestic R&D and innovation systems. Domestic innovation creates domestic technological capacities and capabilities, which increase the potential for technical progress through the interdependent process of domestic knowledge creation and the development of an absorptive capacity: the economic dynamism created by local innovation forms the basis for knowledge assimilation without which foreign technology cannot be absorbed and successful take-off that leads to catching up cannot take place.

The argument is supported by the observations made by economists of technical change regarding the dual role of innovative activities. For example, Cohen and Levinthal (1989:569) argue that "while R&D obviously generates innovations, it also develops the firm's ability to identify, assimilate, and exploit knowledge from the environment." They further qualify this argument and postulate that "firms may conduct basic research less for particular results than to be able to provide themselves with the general background knowledge that would permit them to exploit rapidly useful scientific and technological knowledge...", Cohen and Levinthal (1990:148). Basic research broadens the knowledge base to create a critical overlap with new knowledge. In a similar vein, Abramovitz (1986) suggests that technical congruence is one of the elements that supports the capacity of followers to exploit existing knowledge.

Foreign R&D is often considered as the main means of acquiring technology, and an analysis of north-south spillovers has led to a heated debate. Substantial economic literature propounds that technological growth in developing countries depends on foreign technology acquired through international transfer of technology, and as a result technology diffuses from the north to the south resulting in a reduction of the technology gap over time. For example, Coe *et al.* (1997) empirically examine the extent to which developing countries, which hardly investment in their own R&D benefit from R&D performed in industrialised countries, and conclude that spillovers from the north to the south are substantial. Such contentions have been met with resistance in view of the fact that foreign R&D cannot on its own revamp systems of innovation: it appears unlikely that foreign technology may have much impact in the absence of an absorptive capacity. Indeed, the capacity to benefit from foreign technology appears to depend on the systems of innovation whose development relies largely on domestic innovation rather than on foreign technology.

An avalanche of empirical studies indicating that technology diffusion from industrialised countries has stronger effects in relatively rich countries than in poorer ones reinforces this point (e.g. Eaton and Kortum, 1996; Xu, 2000; Keller, 2001). It is more probable that development of an absorptive capacity - which implies the need to focus on investment in domestic R&D, and human capital development as well as reinforcement of networks and linkages in the case of poorer developing countries - is paramount for productivity growth. The paper shows that domestic innovation lies at the core of the technology gap and is key to shrinking income differences over time.

Traditionally the concept of absorptive capacity has been associated with R&D activities in firms. Recent literature has broadened it to relate to competence building in a rapidly changing economy as well as to include larger entities such as industrial districts, countries and regions. It is noteworthy that innovation arising from R&D is not the autonomous determinant of technical change: incremental transformations are responsible for the bulk of technological knowledge. In the analysis, domestic innovation in developing countries specifically relates to innovative activities based mainly on incremental knowledge. The paper defines variables that relate to innovative activities, and in particular to technological knowledge dynamism at an economy level and then analyses their trends across groups of developing countries. The aim is to map out countries' ability to establish technological learning systems, and hence, to create technological knowledge that leads to technical progress.

The approach that is used consists in viewing total factor produc-tivity as a residual in the production function. The residual is obtained by computing the ratio of national income to factors of production in a model that relates output to factor inputs, and a relationship between total factor productivity growth and both domestic and foreign know-ledge is established in the next section. Section 3 discusses the estimation procedure of our dynamic panel data model and results of the estimation are presented in section 4. The last section concludes.

2. The model

The analysis is based on the approach introduced in the 1950s that views the residual of a Cobb-Douglas aggregate production function as the technology component. The limitations of using a TFP approach are acknowledged in this paper; it does not capture the systemic factors that determine the capacity to create knowledge, and therefore play a large part in driving economies apart. However, owing to the fact that it continues to inspire policy prescriptions, the paper attempts to integrate a variable that captures domestic knowledge, which is generally excluded from an analysis of developing countries despite its importance, to determine whether the significant and positive results obtained in other studies can be confirmed.[2]

$$Y = AF(K,H,L) \dots\dots\dots\dots\dots\dots\dots\dots\dots\dots\dots\dots\dots 1$$

Output Y depends on technology A, physical capital K, human capital H and labour L. One way of increasing output consists in increasing labour and/or investing in physical and human capital. However, growth of output ultimately yields to diminishing returns. The second way requires the improvement of the efficiency with which factor inputs are used, i.e. improving technology A, and it results in sustainable growth.

In his estimates on productivity growth in the US economy, Solow (1957) found that technical change accounted for 80% of per capita growth while capital accumulation accounted for the remaining 20%. Easterly and Levine (2002) also found that technology, other than that incorporated in inputs, plays a fundamental role in growth. Technology or that 'something else' (as they termed it) was found to constitute two thirds of output while inputs accounted for only one third. Our study focuses on this technology term A.

Technological knowledge A is the component that permits countries to take off and maintain sustainable growth because it leads to an increase in output per unit input. Changes in the productivity of production processes are usually measured by variations in total factor productivity, the efficiency with which factor inputs are used. Cross-country differences in total factor productivity reflect differences in technology level. Total factor productivity is thus taken as a measure for the contribution of technical change to growth (Kaldor, 1957).[3]

Measurement of total factor productivity

A production function approach is used to relate total factor productivity to domestic and foreign innovation efforts. A Cobb-Douglas specification for aggregate production appears appropriate in the determination of total factor productivity since the rates of return to factor inputs form constant proportions of national income over time, which is one of stylised facts of economic growth (Kaldor, 1961).

Mankiw *et al.* (1992) integrate human capital in the textbook Solow growth model, which assumes a Cobb-Douglas production function. The resulting so-called "Augmented Solow model" takes the time spent in school as a measure of human capital investment. However, their integration of schooling in the Cobb-Douglas specification for aggregate production has a drawback: the rate of return to schooling is inversely proportional to years of schooling in the workforce, consequently implying high returns to schooling in countries with low stocks of education. Bloom *et al.* (2004) note that in microeconomic studies, returns to education are found to be constant across countries, but no systematic variations of returns to schooling with income or years of schooling of the workforce are observed.

The paper uses a standard production function in which aggregate production results from physical capital and human capital adjusted labour inputs,

$$Y_{it} = AK_{it}^{\alpha}\left(e^{\phi s}L_{it}\right)^{1-\alpha} \quad \dots\dots\dots\dots\dots\dots\dots\dots\dots\dots\dots\dots\dots\dots\dots\dots\dots\dots\dots 2$$

where $e^{\phi s}L_{it} = hL_{it} = H_{it}$

Y is the output, A is technology, K is physical capital, H is human capital (skilled labour) which is produced from raw labour (unskilled labour) L by means of education, and where s represents the average time spent in school (it is the ratio of total time spent in school to total labour force and is taken to be a proxy for human capital investment), while ϕ is the natural rate of return to schooling. Human capital is a simple Mincerian function of schooling.[4] The subscripts i and t denote country and time respectively.

The parameters of the production function are represented by α and $(1-\alpha)$. Each factor earns its marginal product so that α is the share of national income that goes to capital while $(1-\alpha)$ is the share of national income that goes to wages of the labour force. The total wage payments $(1-\alpha)Y$ do not distinguish between returns to raw labour and returns to schooling. The marginal product of an extra year of schooling is ϕY while the marginal product of a worker is $\dfrac{(1-\alpha)Y}{L}$.

In the analysis, it is assumed that an extra year of schooling adds proportionately to output regardless of the level of schooling of the worker obtaining an extra year of schooling.[5] The marginal benefit of an extra year of schooling is the same for all workers regardless of the time spent in school by an individual worker.[6]

The log of output per labour unit i depends on log capital per worker (capital intensity) plus log of human capital intensity and other factors captured in the residual. Dividing both sides of the specified aggregate production function by labour, taking the logs and dropping the indices for simplicity yields,

$$\log(Y/L) = \log A + \alpha \log(K/L) + (1-\alpha)\log\left(e^{\phi s}L/L\right) \quad \text{................................} \quad 3$$

Extracting total factor productivity and using lower case notation to indicate logs yields,

$$p_{it} = y_{it} - \alpha k_{it} - (1-\alpha)\phi s_{it} \quad \text{..} \quad 4$$

where $\log A$ is represented by p_{it}.

Analysis of total factor productivity growth

The analysis relates total factor productivity to both foreign and domestic knowledge. This production function approach is one of the main methods used in analysing the impact of foreign knowledge on domestic productivity in a regression framework. Economic literature identifies four sources that contribute to the improvement of productivity; domestic sources on the one hand that include domestic R&D and outward FDI, and foreign sources on the other hand which are made up of foreign R&D (via imports and partnerships/licensing) and inward FDI.

Improvement of total factor productivity is a process that results from learning and innovation efforts of both domestic and foreign firms. As noted earlier, innovation efforts by domestic firms lead to the creation of an absorptive capacity without which foreign technology is not likely to benefit domestic economies. The term absorptive capacity is used to refer to the ability to improve productivity through the adoption and application of foreign knowledge. Thus, domestic innovative efforts boost the learning capability that is critical for take-off and subsequent catch-up, which requires foreign knowledge.

In the absence of domestic sources of knowledge, particularly domestic innovation, which normally precedes outward FDI, direct attempts to inject foreign knowledge (through, for example, high-technology content goods) are bound to penalise the learning process that leads to knowledge accumulation by provoking a fall in labour productivity. Furthermore, to a large extent foreign knowledge is induced by the presence of an absorptive capacity: the absence of an absorptive capacity, which reflects a weak learning process, inhibits foreign knowledge diffusion into domestic economies.

The implication here is that omission of domestic sources of knowledge from the estimation, as is often the case in empirical studies dealing with developing countries whose domestic innovation efforts are feeble while outward FDI is practically non-existent, may lead to bias of estimates as discussed later in more detail.

(a) Foreign R&D

It is assumed that foreign knowledge resulting from R&D efforts is transmitted to developing countries through imports of high technology content capital goods. v_{it}^M captures the real R&D intensity embodied in imports following Lichtenberg and van Pottelsberghe de la Potterie (1996). An argument is put forward regarding the effect of foreign R&D capital stock on developing countries as occurring primarily and perhaps entirely through the indirect channel of trade since licensing/partnerships occur almost exclusively amongst industrialised countries. Thus foreign R&D capital stock of a country i is represented by,

$$v_i^M = \sum_j \left(v_j^d / y_j\right) m_{ij} \dots\dots\dots\dots\dots\dots\dots\dots\dots\dots\dots\dots\dots\dots\dots\dots\dots\dots\dots 5$$

where i and j represent the developing country and the industrialised country indexes respectively, v_j^d represents the domestic R&D capital stock of the industrialised country j, m_{ij} is the total imports of the developing country i from the industrialised country j, and y_j represents the GDP of the industrialised country j. The R&D intensity in the industrialised country is represented by v_j^d / y_j, but since the same group of industrialised is used as the trade partners for developing countries, the R&D intensity of industrialised countries is a constant term that may be eliminated from the equation.[7]

(b) Inward FDI

Foreign knowledge embodied in inward FDI is computed to capture the intensity of foreign R&D in inward FDI. Thus,

$$v_i^{FDI} = \sum_j s_{ij} \left(v_j^d / k_j\right) \dots\dots\dots\dots\dots\dots\dots\dots\dots\dots\dots\dots\dots\dots\dots\dots\dots\dots\dots\dots 6$$

where s_{ij} is the inward FDI flows of the developing country i emanating from the industrialised country j, while v_j^d represents the domestic R&D capital stock of industrialised country j, and k_j is the capital stock of the industrialised country j. The R&D intensity of capital stock of industrialised countries may be interpreted as a constant because the same group of industrialised countries is maintained. It is therefore, eliminated from the equation.

(c) Domestic knowledge

While domestic innovation via both domestic R&D and outward FDI, has been found to play a critical role in productivity growth, particularly with regard to studies on industrialised countries, most empirical studies on developing countries do not account for it. The argument put forward is that developing countries' domestic innovation is insignificant and worse still, data is unavailable. Although this argument may be some-what valid, it is considered that the inclusion of a variable in the estimation specification reflecting the insignificance of domestic inno-vation is crucial.

To the extent that domestic innovation creates technological know-ledge that is instrumental in the initial creation of an absorptive capacity, which has been identified as the element responsible for take-off and catch-up, it is important to identify a variable that relates to the absorp-tive capacity. Such a variable would enable us to gain some understand-ing of why some countries are unable to take-off, and in some cases recede further into marginalisation.

It is noted that building-up of the learning capability, which allows the creation of an absorptive capacity, must take place during the pre-catching-up phase if take-off is expected to occur; as suggested by Cohen and Levinthal (1990) prior knowledge, which at the most elementary level includes basic skills, is the foundation for the 'initial' absorptive capacity. It is assumed therefore, that learning capability fundamentally determines the creation and development of an initial stock of knowledge that triggers the cumulative and interactive process between knowledge stock and absorptive capacity, and thus sparks take-off.

In more general terms, the creation of a prior technological know-ledge is closely tied to human capital development. Creation of know-ledge arises from a variety of sources such as formal education, vocational training, in-firm training, learning on the job, and specialised employee training outside the firm (Lall, 2000). The nature of formal education and vocational training in the economy determines the level of sophistication in the technologies employed. Modern technology requires fairly high levels and broad coverage of formal education and training. Hence, in-firm training, on the job learning, and specialised employee training outside the firm are calibrated on the base of formal education and training available in the economy.

Indeed, economic literature argues that human capital contributes to production directly (marginal product) and indirectly by inducing foreign knowledge - via capital imports of high technology contents, inward FDI, and licensing (in the case of industrialised countries) - and facilitating its use resulting in enhanced productivity growth. The indirect mechanism relies on competence creation, which occurs via domestic innovation. Domestic innovation is knowledge intensive and, hence, thrives upon human capital (Romer,1990). The productivity enhancing effect of human capital is increasingly identified as the link between education and growth: education policies oriented towards requirements in the business sector play a determinant role in economic performance. Hence, human capital is critical in an estimation specification explaining productivity growth.

A term relating the effects of human capital on productivity with the technological distance from the frontier appears relevant to the estimation specification. Since the interest lies in the indirect rather than direct effect of human capital on productivity in the definition of this variable, it is perhaps more interesting to interact it with a term that relates to the efficiency level, which may be referred to as the distance to the technological frontier: an estimation specification with an interaction coefficient may provide more accurate results.

The distance from the technological frontier, or backwardness, may be viewed as the efficiency level of a country, which reflects the "quality" of the innovation system defined to include economic, social and political infrastructures and institutions. This would probably give a more accurate specification and perhaps remedy the problem of variable omission that ultimately leads to bias of estimates. Measurement of the "quality" of the innovation system or the efficiency level is a major concern.

One way in which empirical literature resolves this measurement problem consists in using the GDP ratio of machinery equipment imports to reflect the technological distance of a country from the frontier (Mayer, 2001; Coe *et al.* 1997). It is noted that the term obtained from interacting the GDP ratio of machinery equipment imports with human capital mirrors to some extent the absorptive capacity of country: the larger the ratio, the greater the indirect effect of human capital, which implies a greater capacity to reach the technology frontier through the cumulative and interactive process between knowledge stock and the capacity to assimilate foreign knowledge.

3. Estimation[8]

Our estimation specification is defined as a state dependent model,

$$p_{it} = \varphi p_{it-1} + \beta_1 v_{it}^M + \beta_2 v_{it}^{FDI} + \beta_3 v_{it}^D + \lambda_t + \mu_i + \omega_{it} \dots\dots\dots\dots\dots\dots\dots 7$$

where the total factor productivity is denoted by p_{it}, the lagged dependent variable by p_{it-1}, foreign R&D by v_{it}^M, inward FDI by v_{it}^{FDI}, and domestic knowledge v_{it}^D. Ideally, the domestic knowledge variable should be represented by domestic R&D and outward FDI. It is assumed that developing countries do engage in these two activities in one way or another, but data is not available for the whole sample[9]. Therefore, in the estimation domestic knowledge v_{it}^D is replaced with an interaction term between human capital and the efficiency of production (GDP ratio of machinery and equipment imports as a proxy of production efficiency).[10] The country specific variable (representing for example geography) is denoted by μ_i, and λ_t denotes a time effect (captures the effect of the time variant technology frontier) such that $\lambda_{it} = \lambda_t + v_{it}$ where v_{it} is included in the error term ω_{it}.

Path dependence is a major factor influencing technology acquisition: it appears reasonable to assume that past productivity p_{it-1} influences current productivity p_{it}. In addition, past productivity may influence the other explanatory variables as discussed in the next subsection in greater detail. A dynamic model appears appropriate.

The standard methods that are used to estimate panel data models are fixed effects or random effects with the major difference between the two being the information utilised to calculate the coefficients: the fixed effect estimates are calculated from differences within each country across time and the method does not account for the presence of unobserved time invariant characteristics (it simply absorbs them into the fixed effects), while the random effects estimates incorporate information across individual countries as well as across periods. Although the random effects estimates may be more efficient, the method requires that the country specific effects be uncorrelated with the explanatory variables for estimates to be consistent which is often unlikely. A Hausman specification test may to some extent be used to evaluate whether this independence assumption is satisfied.

Hausman and Taylor (1981) propose the use of an instrumental variables' estimation as a way to overcome the problem of bias in the estimates. Their approach entails transformation of the model to deviations from county means in order to get rid of the country specific effects that are correlated with the explanatory variables. The country mean deviations are used as instrumental variables to obtain consistent estimators.

However, even though the instrumental variable estimator is consistent it may not be efficient; correlation between the explanatory variables and the disturbance may still exist. Furthermore, the presence of a lagged dependent variable in our model makes the Hausman & Taylor approach inappropriate as it is not directly applicable to a dynamic model: the presence of a lagged dependent variable in the model violates the assumption of strict exogeneity because the lagged endogenous variable is bound to be correlated with the error term. In addition, since the time series dimension is fixed ($t = 21$ or $t = 5$ i.e. t does not approach infinity), the estimation is not consistent even as n goes to infinity. Hence, the bias for the coefficient of the lagged endogenous variable may be significant.

Arellano and Bond (1991) suggest an alternative estimation technique that corrects for the bias introduced by the lagged endogenous variable, and in addition, permits a certain degree of endogeneity in other regressors. A more detailed discussion of the model is presented below.

Effects of the absorptive capacity on productivity growth

The growth of total factor productivity is examined using a sample of 51 developing countries over the period 1981- 2000.[11] The productivity growth equation:

$$p_{it} - p_{it-1} = (\varphi - 1)p_{it-1} + \beta_1 v_{it}^M + \beta_2 v_{it}^{FDI} + \beta_3 v_{it}^D + \lambda_t + \mu_i + \omega_{it} \dots\dots\dots 8$$

may be rewritten as:

$$p_{it} = \varphi p_{it-1} + \beta_1 v_{it}^M + \beta_2 v_{it}^{FDI} + \beta_3 v_{it}^D + \lambda_t + \mu_i + \omega_{it} \dots\dots\dots 9$$

The model holds for the years 1981 to 2000 with p_{i0} corresponding to 1980, the first year of data. It is assumed that one lag of the dependent variable, p_{it-1} is sufficient to capture the dynamics in the conditional expectation and any further lags on p_{it} or lags on the other explanatory variables are unimportant (the inclusion of p_{it-1} in the model along with other explanatory variables is intended to control for another source of omitted variable bias). The value of φ need not be restricted given that the analysis is based on fixed time asymptotics. The coefficient of interest is on the domestic knowledge indicator v_{it}^D which captures the absorptive capacity of a country. A robust and positive β_3 is expected.

One implication of the above model is that the lagged dependent variable is correlated with the disturbance (even if it is assumed that the disturbance itself is not auto-correlated) because of a possible bias by the individual specific effects since the same specific effect enters the equation for every observation in each group. $E(\omega_{it}|p_{it-1}) \neq 0 \quad t = 2,3,...T$ and an estimation of the model using the usual techniques would lead to an inconsistent estimator. Arellano and Bond propose an alternative estimation technique that corrects the bias introduced by the lagged dependent variable. The idea consists in first differencing the productivity growth equation,

$$p_{it} - p_{it-1} = \varphi(p_{it-1} - p_{it-2}) + \beta_1(v_{it}^M - v_{it-1}^M) + \beta_2(v_{it}^{FDI} - v_{it-1}^{FDI})$$
$$+ \beta_3(v_{it}^D - v_{it-1}^D) + (\lambda_t - \lambda_{t-1}) + (\omega_{it} - \omega_{it-1}) \quad\quad\quad\quad \text{10}$$

Equivalently,

$$\Delta p_{it} = \theta_t + \varphi \Delta p_{it-1} + \beta_1 \Delta v_{it}^M + \beta_2 \Delta v_{it}^{FDI} + \beta_3 \Delta v_{it}^D + \Delta \omega_{it} \quad\quad\quad\quad \text{11}$$

where time dummies are represented by $\theta_t = \lambda_t - \lambda_{t-1}$.

The first differencing transformation eliminates the country dummies (unobserved country effects) μ_i, and thus the bias introduced by the lagged dependent variable, and therefore allows the use of a simple instrumental variable estimator.[12] However, correlation between the lagged dependent variable and the disturbance still exists since past productivity influences the current level of foreign R&D spillovers, inward FDI and domestic knowledge: $V_{it} = \xi p_{it-1} + \alpha_t + \phi \mu_i + \varepsilon_{it}$ where $V_{it} \equiv (v_{it}^M, v_{it}^{FDI}, v_{it}^D)$. Lagged values of each of the independent variables are used as instruments so as to remedy the correlation problem between the explanatory variables and the disturbance $E(\omega_{it}|V_{it}) \neq 0 \quad t = 1,2,3,...T$.

$$E(V_{it}\omega_{is}) \begin{cases} \neq 0 & s < t \\ = 0 & s > t \end{cases} \quad\quad V_{it} \text{ is predetermined and not strictly exogenous.}[13]$$

4. Results

Table 1 below reports estimates of the productivity growth equation using fixed effects, random effects, Hausman & Taylor procedure and the Arellano & Bond GMM technique. Estimates vary depending on the technique that is used, making it necessary to test the validity of the assumptions underlying each method. First a Hausman specification test comparing the fixed-effects estimates in column [1] with the random effects in column [2] rejects the assumption that country specific effects are uncorrelated with the explanatory variables as is required for random effects. Nonetheless, both methods are inconsistent due to the presence of the lagged endogenous variable.

The coefficients of the Hausman & Taylor estimator reported in column [3] are virtually similar to those obtained by the fixed effects estimator in column [1] suggesting that specific effects do not bias the model and should therefore be included in the estimation equation. However, coefficients of the lagged dependent variables obtained by the Arellano & Bond approach used in column [4] are large and highly significant, suggesting that this method is preferable to the Hausman & Taylor technique used in column [3] whose estimates are inconsistent because it is also a static model (it does not take into account the lagged dependent variable). This is an informal way of selecting between the static and dynamic model since no formal test exists. The presence of a lagged dependent variable points, *a fortiori*, to a dynamic rather than a static model. It is noteworthy that had the coefficients obtained in column [4] not been robust this would have indicated the need to perhaps redefine the estimation specification; a state dependent model would not have been appropriate.

Estimates obtained using lagged instruments of the explanatory variables or regression of explanatory variables on the lagged dependent variable, suggest that past productivity influences the current level of the explanatory variables. For example, regressing foreign R&D spillovers on the lagged dependent variable suggests that past productivity influences the current level of foreign R&D spillovers, i.e. $v_{it}^M = \xi\, p_{it-1} + \alpha_t + \phi\, \mu_i + \varepsilon_{it}$. This implies that $V_{it} \equiv \left(v_{it}^M, v_{it}^{FDI}, v_{it}^D\right)$ are predetermined by at least one period. Although endogeneity may exist between knowledge variables $V_{it} \equiv \left(v_{it}^M, v_{it}^{FDI}, v_{it}^D\right)$ and productivity growth, the test for autocorrelation and the Sargan test of over-identifying restrictions satisfy the underlying assumptions of the Arellano & Bond approach suggesting that estimates reported in column [4] are consistent and efficient.[14]

Table 1: Regression results

REGRESSION RESULTS: ALTERNATE ESTIMATION TECHNIQUES

estimation method	Fixed effects	Random effects	Hausman & Taylor	Arellano & Bond	Arellano & Bond diff gmm	Arellano & Bond system gmm	Arellano & Bond system gmm
			period 1980-2000				5 five-year periods
	[1]	[2]	[3]	[4]	[5]	[6]	[7]
	pdvty	pdvty	pdvty	D.pdvty	pdvty	pdvty	pdvty
fkm	0.08	0.078	0.08		0.087	0.032	0.056
	(7.39)**	(7.45)**	(7.37)**		(7.92)**	(9.49)**	(3.08)**
flkfdi	-0.019	-0.018	-0.019		0.031	-0.002	0.025
	(3.28)**	(3.12)**	(3.27)**		(5.72)**	(-0.76)	(-2)
dk	-0.064	-0.061	-0.064		-0.008	-0.024	-0.05
	(5.40)**	(5.19)**	(5.39)**		(2.24)*	(7.07)**	(-4.27)**
lpdvty					0.612	0.923	0.589
					(14.14)**	(47.95)**	(7.46)**
const	4.183	4.188	-7.439			0.148	1.35
	(69.37)**	(57.75)**	(-0.03)			(-1.79)	(-4.21)**
LD.pdvty				0.845			
				(14.59)**			
D.fkm				0.05			
				(3.63)**			
D.flkfdi				0.013			
				(-1.23)			
D.dk				-0.008			
				(-0.72)			
ctry			0.383				
			(-0.05)				
obs	1071	1071	1071	969	969	1020	204
countries	51	51	51	51	51	51	51
R-squared	0.06						

Absolute value of t statistics in parentheses
* significant at 5%; ** significant at 1%

The coefficients of both the lagged dependent variable (lpdvty), and the foreign R&D variable (fkm) are positive and highly significant in all estimation techniques as expected. In addition, coefficients of the lagged dependent variable are fairly large, suggesting that past productivity plays a crucial role in future productivity.

The variable representing foreign knowledge via FDI (fkfdi) gives mixed results in columns [4] to [7]. The original Arellano & Bond dynamic panel data estimator in column [4] reports a positive but insignificant coefficient. This result is improved by the Arellano & Bond "difference GMM estimator" in column [5], which is better than the original model. It provides a finite sample correction to the two-step covariance matrix that compensates for the severely down biased two-step estimates of the standard error, obtained in the original model. However, lagged levels in both the original Arellano & Bond estimator as well as the "difference GMM estimator" are usually poor instruments for the first differences, and especially for variables which are close to a random walk, which is the case in the explanatory variables of the model, and are therefore probably biased.

Indeed, the Arellano & Bond "system GMM estimator" in column [6], which is an augmented version of the "difference GMM estimator", does not confirm the result in column [5]. In the augmented version, original equations in levels are added so as to provide additional moment conditions that are used to increase the efficiency of the estimates. The "system GMM estimator" reports a negative coefficient, but it is not significant. In a further step, "more developed" developing countries are removed from the regression and an estimation is carried out for 5 five-year periods, which implies that $t = 5$ instead of $t = 21$.[15] This mitigates the problem of loss of degrees of freedom. A negative and highly significant coefficient for foreign knowledge via FDI is obtained from the "system GMM estimator". A similar regression is carried out for the "more developed" developing countries. A positive and significant coefficient is obtained for this group of countries. These results appear particularly interesting and lead to the conclusion that potential benefits of FDI accrue only to the small group of "more developed" developing countries that engage in domestic investment and thus, dispose of a relative absorptive capacity. Indeed, these are also the countries that would be able to attract market seeking FDI (horizontal FDI that is more pervasive in introducing foreign knowledge than vertical FDI), rather than serve as mere export platforms (vertical FDI).

This finally brings us to the coefficient of main interest, domestic knowledge (dk), which gives the expected result: the coefficient is negative and highly significant in all estimation techniques except for the original Arellano & Bond estimator in column [4], which reports a non significant coefficient, while the Arellano & Bond "difference GMM estimator" reports a significance level of 5%. One interesting observation is that the coefficient remains negative throughout; it supports the initial view that although the commonly used interaction coefficient between human capital and the GDP ratio of high technology imports may to some extent depict the absorptive capacity of a country it mainly portrays openness of an economy. This may lead to the conclusion that opening up fragile economies is likely to result in a negative effect on the productivity growth of these economies. Although economic research on the role of openness in developing countries has led to mixed results, a number of interesting papers including Fagerberg and Verspagen (2004) find that opening up weak economies is bad for growth.

5. Conclusions

The results support the view that foreign knowledge generates a beneficial impact on the economic performance of the few developing countries that have been successful in embarking on an innovation-driven growth path by simultaneously engaging in technical competence creation and innovation. This is particularly evident for foreign knowledge via FDI. With regard to foreign R&D, the results suggest positive and highly significant benefits for the whole sample. However, it may be argued that the calculation of the foreign R&D variable is based on imports of high technology content capital and machinery, which mainly capture openness and do not necessarily suggest significant spillovers. More specifically, efforts to inject foreign knowledge through high technology content imports in weak economies are bound to penalise the learning process that leads to knowledge accumulation by provoking a fall in labour productivity. Devarajan *et al.*'s (2001) study on sub-Saharan Africa revealed that an increase in capital accumulation in Tanzania led to a fall in output per unit of labour and consequently a fall in output per unit of capital due to underutilisation.

Estimation of the dynamic panel data model using alternative methods, for example the Kalman filter, to infer data for domestic knowledge may have provided an interesting basis for comparison with estimations that use the interaction coefficient between human capital and the GDP ratio of high technology imports as a proxy for domestic knowledge.

However, it is noted that the explanatory power of linear models may be quite limited as concerns the absorptive capacity. The absorptive capacity is most probably represented by a sigmoid function, a functional form that approximates the stylised S-shaped function of technology diffusion models. A nonlinear logistic specification is much more likely to be robust. Benhabib and Spiegel's (2002) estimation of a logistic specification reveals that divergence is a possible outcome for countries with no absorptive capacity.

To the extent that a solid technological infrastructure is indispensable for sustained growth, and that investment in knowledge producing activities may be scarce as is the case in most developing countries, there is a rationale for public intervention with strong policy co-ordination that favours technological shifts. Admittedly, limited innovation may be caused by such factors as inadequate environment for risk taking, unavailability of information about technological opportunities, inadequate inputs (particularly competences), and taxation systems that fail to induce industrial activities.

Appendix 1

Country sample

Country list of 51 developing countries used in the analysis
(definition of developing countries is that of the WTO)

Africa (21 countries)	*Latin America (17 countries)*	*Asia (13 countries)*
Algeria	Argentina	Bangladesh
Benin	Bolivia	China
Cameroon	Brayil	Honk Kong
Central African Republic	Chile	India
Congo, Dem. Rep.	Colombia	Indonesia
Congo, Republic of	Costa Rica	Korea, Republic of
Egypt	Ecuador	Malaysia
Ghana	El Salvador	Nepal
Kenya	Guatemala	Pakistan
Malawi	Honduras	Philippines
Mali	Mexico	Singapore
Mauritius	Nicaragua	Sri Lanka
Mozambique	Panama	Thailand
Niger	Paraguay	
Rwanda	Peru	
Senegal	Uruguay	
Togo	Venezuela	
Tunisia		
Uganda		
Zambia		
Zimbabwe		

"more developed" developing countries *(group one countries)*	*Other developing countries* *(group two countries)*	
Argentina	Algeria	Mauritius
Brazil	Bangladesh	Mozambique
Chile	Benin	Nepal
China	Bolivia	Nicaragua
Egypt	Cameroon	Niger
Hong Kong	Central African Republic	Panama
India	Colombia	Paraguay
Indonesia	Congo, Dem. Rep.	Peru
Korea, Republic of	Congo, Republic of	Philippines
Mexico	Costa Rica	Rwanda
Pakistan	Ecuador	Senegal
Singapore	El Salvador	Sri Lanka
Thailand	Ghana	Togo
Venezuela	Guatemala	Tunisia
	Honduras	Uganda
	Kenya	Uruguay
	Malawi	Zambia
	Malaysia	Zimbabwe

A sample of 22 advanced countries used in the analysis: Austria, Belgium, Denmark, Finland, France, Germany, Greece, Italy, Ireland, Luxembourg, Netherlands, Portugal, Spain, Sweden, United Kingdom, Australia, Canada, Japan, New Zealand, Norway, Switzerland, United States of America.

Data

Capital stock

The initial physical capital stocks are calculated using the method proposed by Klenow and Rodriguez-Clare (1997)[16]

$$\frac{K}{Y}_{1980} = \frac{I_K/Y}{g+d+n} \quad\dots\dots\dots\dots\dots\dots\dots\dots\dots\dots\dots\dots\dots\dots\dots\dots\dots\dots\dots\dots\dots\dots\dots\dots\dots\dots\dots 1$$

where I_K/Y is the average investment rate in physical capital (1980-2000), g is an estimation of the world average growth rate of output per capita Y/L given as 0.02, d represents the rate of depreciation which is set at 0.03, and n is the rate of growth of the working population 15-64 year olds (1980-2000). The depreciation rate is taken from Mankiw *et al.* (1992) based on a calculation for a large sample of countries. Although the depreciation rate for fast growing developing countries may vary widely, data that would allow the estimation for country-specific depreciation rates for the whole sample is not available. The physical capital stock of a country i in period t satisfies as in Benhabib and Spiegel (1994) the equation:

$$K_{it} = \sum_{\varepsilon=0}^{t}(1-d)^{t-\varepsilon}I_{i\varepsilon} + (1-d)^{t}K_{1980} \quad\dots\dots\dots\dots\dots\dots\dots\dots\dots\dots\dots\dots\dots\dots 2$$

Data for real income (PPP GDP), employment/labour (population) and PPP investment in physical capital are from the Penn World Tables version 6.1 (2002). Data for schooling, which is given as the average years of schooling in the population above 25 years of age, is obtained from Barro and Lee's data set (2000). The constant marginal rate of return to physical capital is set at $\alpha = 1/3$. The assumption of $\alpha = 1/3$ is based on Bernanke and Gurkaynak's (2001) calculation of the labour share of 0.65 and by implication a capital share of 0.35[17]. The rate of return to schooling $\phi = 0.085$.

Foreign R&D

Data on machinery and transport equipment is obtained from the UN Comtrade database section 7 of SITC Rev. 2 from which consumption goods as well as parts and components imported by developing countries for re-export after incorporating some form of value added are omitted. The analysis is based on mirror trade data: imports by developing countries are assumed to be equivalent to exports by partner (industrialised) countries, due to the unavailability and unreliability of import data of most developing countries. The breakdown is as follows:

Machinery: SITC Rev. 2: 71-77 less 761-3, less 775-776
Transport & Equipment: SITC Rev. 2: 78-79

Inward FDI

Data is from UNCTAD Foreign Direct Investment Database (2004) The database presents aggregate inward FDI stocks. It is assumed that the inward FDI stocks in developing countries emanating from the world rather than from the selected group of industrialised countries does not significantly alter results. In addition, it is noted that inward FDI may not constitute a significant channel through which knowledge is diffused: inward FDI may not contribute to the improvement of the host country's productivity since the foreign owner has no incentive to share technology and may prefer to adapt to the host country's technology. Indeed, inward FDI typically takes place via a wholly owned subsidiary in a bid to keep technology under the control of the multinational.[18]

Domestic knowledge

The domestic knowledge variable of a country is defined as an interaction term between human capital and the GDP ratio of machinery equipment imports. The GDP ratio of machinery equipment imports is calculated using the UN Comtrade database section 7 of SITC Rev. 2 as described above and GDP from Penn World Tables. The human capital data is obtained from Barro and Lee (2000) who generate a comparable estimate (h) for a large sample of countries. The underlying condition in this approach, however, is that there exists an adequate level of human capital, which brings us back to the importance of building what was referred to as a learning capability. In other words, direct attempts to inject foreign knowledge in economies that are poorly endowed in human capital may penalise the learning process that leads to knowledge accumulation by provoking a fall in labour productivity.

Appendix 2

Determination of instruments

The instruments are determined as follows:

For the period $t = 3$ the productivity equation may be written as:

$$p_{i3} - p_{i2} = \varphi(p_{i2} - p_{i1}) + \beta(V_{i3} - V_{i2}) + \theta_t + (\omega_{i3} - \omega_{i2}) \dots\dots\dots\dots\dots 1$$

In the third period p_{i1} may serve as an instrument since it is highly correlated with $(p_{i2} - p_{i1})$, but uncorrelated with $(\omega_{i3} - \omega_{i2})$ if ω_{it} is a white noise. As for $(V_{i3} - V_{i2})$, V_{i1} and V_{i2} are valid instruments since they are not correlated with the error term $(\omega_{i3} - \omega_{i2})$. [Level instruments are preferable to difference instruments. Orthogonality conditions are stated in terms of the levels of the variables and the differences of the disturbances $E(V_{is}\Delta\omega_{it}) = 0$ as opposed to differences of both the variables and the disturbances $E(\Delta V_{is}\Delta\omega_{it}) = 0$ which is implied $s = 1,...,t-2$, Arellano (1989).

The matrix of instruments may be written as:

$$Z_i = \begin{bmatrix} p_{i1}, V_{i1}, V_{i2} & d_{1982} & \cdots & \cdots & 0 & 0 \\ & p_{i1}, p_{i2}, V_{i1}, V_{i2}, V_{i3} & d_{1983} & & \vdots & \vdots \\ \vdots & \vdots & & \ddots & & \\ 0 & 0 & \cdots & \cdots & p_{i1},...p_{iT-2}, V_{i1},...V_{iT-1} & d_{2000} \end{bmatrix}$$

d_{year} represents the year specific dummy variable.

Once the instruments are identified the instrumental variables method is applied to the first differenced productivity equation

$$\Delta p_{it} = \theta_t + \varphi \Delta p_{it-1} + \beta \Delta V_{it} + \Delta \omega_{it} \dots\dots\dots\dots\dots\dots\dots 2$$

Let $\begin{bmatrix} \varphi \\ \beta \end{bmatrix} = \delta$. A convergent estimator of the parameter δ is obtained but, the GMM estimator of δ may not be efficient since (ω_{it}) is a random walk with a unit root: $\omega_{it} - \omega_{it-1} = \Delta\omega_{it}$ hence, $\omega_{it} = \omega_{it-1} + \Delta\omega_{it}$ is a random walk since it is assumed that $\Delta\omega_{it}$ has no serial correlation (it is a white noise).

It was assumed that the first difference of the idiosyncratic errors $\Delta\omega_{it} : t = 2,3,...T$ are serially uncorrelated and have constant variance.

$$E(w_i w_i' | p_{i0},...p_{iT-1}, V_{i1},...V_{iT}, \mu_i) = \sigma_w^2 I_{T-1}$$

where w_i is the $(T-1) \times 1$ vector containing $\Delta \omega_{it} : t = 2,3,...T$.

Since this assumption may not be verified $E(w_i w_i') \neq \sigma_w^2 I_{T-1}$ a matrix of instruments $Z = [Z_1',...Z_N']'$ is used such that the orthogonality conditions are now $E(Z_i' \Delta \omega_i) = 0$. This is the weakest assumption that can be imposed in a regression framework to get a consistent estimator of δ. Under this assumption the vector δ satisfies

$$E[Z'\Delta p - Z'(\Delta p_{-1}\Delta V)\delta] = E[Z'\Delta \omega] = 0$$

or equivalently

$$E[Z'(\Delta p_{-1}\Delta V)\delta] = E[Z'\Delta p]$$

where $Z'\Delta p$ is a $K \times 1$ random vector and $Z'(\Delta p_{-1}\Delta V)$ is a $K \times K$. To be able to estimate δ, it is assumed that it is the only $K \times 1$ vector that satisfies the orthogonality condition. This implies that although this orthogonality condition is the basis for estimating δ, the rank condition is required as a sufficient assumption for identification.

The assumption of a full rank implies that the system has a unique solution – there is no over identification

$$rank\left(\sum_{t=2}^{T} E(\Delta X_{it}' \Delta X_{it}) \right) = K$$

$$\Delta X_{it} \equiv (\Delta p_{it-1}, \Delta V_{it})$$

Time constant explanatory variables and perfect collinearity among the time varying variables is ruled out. The matrix is non-singular, which rules out the presence of linear dependence.

Estimating δ

With the orthogonality conditions and the full rank assumption solving, for δ will yield a unique solution. A weighting matrix $\hat{W}$, a positive semi definite matrix, in the quadratic form to obtain $\hat{\delta}$ is used.

$$\hat{\delta} = \min_{\delta} \left[\sum_{i=1}^{N} Z_i'(\Delta p_i - \Delta X_i \delta) \right]' \hat{W} \left[\sum_{i=1}^{N} Z_i'(\Delta p_i - \Delta X_i \delta) \right]$$

Hence,

$$\hat{\delta} = \frac{\Delta p Z' \hat{W} Z' \Delta p}{(\Delta p_{-1} \Delta V) Z' \hat{W} Z'(\Delta p_{-1} \Delta V)} = \left[(\Delta p_{-1} \Delta V) Z' \hat{W} Z'(\Delta p_{-1} \Delta V)\right]^{-1} \left[\Delta p Z' \hat{W} Z' \Delta p\right]$$

First step

The first choice of the weighting matrix $\hat{W}$ is,

$\hat{W} = \left(N^{-1} \sum_{i=1}^{N} Z_i' Z_i\right)^{-1}$ which is a consistent estimator of $\left[E(Z_i' Z_i)\right]^{-1}$

The IV estimator of δ may be written as:

$$\hat{\delta}_{IV} = \left(\left[\Delta p_{-1} \Delta V\right]' Z(Z_i' Z_i)^{-1} Z' \left[\Delta p_{-1} \Delta V\right]\right)^{-1} \left(\left[\Delta p_{-1} \Delta V\right]' Z(Z_i' Z_i)^{-1} Z' \Delta p\right) \quad \dots\dots\dots 3$$

The weighting matrix $\hat{W} = \left(\dfrac{1}{N} \sum_{i=1}^{N} Z_i' Z_i\right)^{-1}$ gives the initial consistent estimator $\hat{\delta}_1$, but may not be necessarily the asymptotically efficient estimator. However, it is important because a preliminary consistent estimator of δ is required to obtain the asymptotically efficient estimator.

Second step

The optimal weighting matrix that produces the GMM estimator with the smallest asymptotic variance is,

$$\hat{W} = \left(\frac{1}{N} \sum_{i=1}^{N} Z_i' \Delta\hat{\omega} \, \Delta\hat{\omega}' Z_i\right)^{-1}$$

The optimal GMM estimator of δ may be written as:

$$\hat{\delta}_{GMM} = \left(\left[\Delta p_{-1} \Delta V\right]' Z \hat{W} Z' \left[\Delta p_{-1} \Delta V\right]\right)^{-1} \left(\left[\Delta p_{-1} \Delta V\right]' Z \hat{W} Z' \Delta p\right) \quad \dots\dots\dots\dots\dots 4$$

Columns [8] and [9] distinguish between "more developed" (group one) and "less developed" (group two) developing countries

Table 2: Regression Results: Alternate Estimation Techniques

REGRESSION RESULTS: ALTERNATE ESTIMATION TECHNIQUES

estimation method	period 1980-2000						5 five year periods		
	Fixed effects	Random effects	Hausman & Taylor	Arellano & Bond	Arellano & Bond diff gmm	Arellano & Bond system gmm	Arellano & Bond system gmm		
							all countries	group two	group one
	[1]	[2]	[3]	[4]	[5]	[6]	[7]	[8]	[9]
				D.pdvty	pdvty	pdvty	pdvty	pdvty	pdvty
fkrm	-	:	:		0.087 (7.92)**	0.032 (9.49)**	0.056 (3.08)**	0.075 (3.35)**	-0.001 (-0.09)
fkrfdi	-	:	:		0.031 (5.72)**	-0.002 (-0.76)	0.025 (-2)	-0.056 (3.56)**	0.037 (2.25)*
dk	-	:	:		-0.008 (2.24)*	-0.024 (7.07)**	-0.05 (4.27)**	-0.027 (-1.21)	0.003 (-0.5)
grp	-	:	:						
lpdvty					0.612 (14.14)**	0.923 (47.95)**	0.589 (7.46)**	1.19 (11.54)**	0.765 (11.53)**
const		:	:			0.148 (-1.79)	1.35 (4.21)**	-1.076 (2.56)*	0.783 (2.77)*
LD.pdvty				0.845 (14.59)**					
D.fkrm				0.05 (3.63)**					
D.fkrfdi				0.013 (-1.23)					
D.dk				-0.008 (-0.72)					
conti		:	:						
ctry		:	:						
obs	1071	1071	1071	969	969	1020	204	148	56
no. ctries	51	51	51	51	51	51	51	37	14
R-squared	0.06								

Absolute value of t statistics in parentheses   * significant at 5%; ** signi

Notes

[1]   Nelson and Winter (1982) articulated the evolutionary theory of firms and markets, Rosenberg (1983) the chain linked model as an alternative to the linear model and Freeman (1987) empirical findings.

[2]   See for example Coe and Helpman (1997), Mohnen (2001).

[3]   In growth accounting, an index that combines all measurable inputs is estimated and used to measure the rate of growth of national income, i.e. to measure total factor productivity. However, a fundamental difficulty in modelling total factor productivity is that no independent measure for it exists.

[4]   Klenow and Rodriguez-Clare (2004) note that although a more complete Mincerian formulation would include years of experience in addition to schooling, taking experience into account has little effect on aggregate levels and growth rates. We therefore adopt the schooling only view of human capital production.

[5]   The effect of schooling on the wages of an individual has been analysed based on the work of Mincer (1974) where a semi-log equation is used to demonstrate that returns to schooling are constant across countries:

$\log W_j = \alpha_0 + \alpha_1 S_j$ where W_j is the wage of an individual j, and S_j his years of schooling. An extra year of schooling increases wages by the amount $\alpha_1 W_j$. The rate of return to schooling α_1 is taken to be the same for each worker regardless of the time spent in school by the individual. The wage equation suggests that returns to uneducated workers α_0 do not depend on the level of schooling in the workforce (Bloom *et al*, 2004). This problem does not appear in the aggregate production function proposed as it is specified in such a way that the wage of an uneducated worker depends on the average level of education.

[6]   However, the formulation of aggregate production that we adopt may imply that the rate of return to schooling is equivalent to the social rate of return of a worker. Bloom *et al.* (2004) propose the formulation $Y_{it} = AK_{it}^{\alpha} e^{\phi s} L_{it}^{1-\alpha}$ where $e^{\phi s} L_{it} = H_{it}$, which implies that the social rate of return for an average worker is $\dfrac{\phi}{(1-\alpha)}$. It can be demonstrated, nonetheless, that different workers face different social rates of return to schooling, $\dfrac{\phi}{(1-\alpha)+\phi(s_j - s)}$ where S_j is the number of years of schooling of a worker j, while S is the average years of schooling in the labour force. This is a limitation that occurs in any aggregate production that depends only on average (total) years of schooling as it makes the assumption that the margi-

nal benefit is the same for all workers regardless of level of schooling, while the cost of schooling takes into consideration the education level of each worker in determining the output forgone by withdrawing a worker from the labour force. An aggregate production function that maintains the Mincer equation property, that the rate of return to education is the same for all workers should include distribution as well as the average level of human capital. In the interest of simplicity, we follow Bloom *et al.* (2004) and assume that only the total stock of education matters and not its distribution.

[7] The group of industrialised countries is indicated in Appendix 1.

[8] See Appendix 1 for data sources, data analysis and computation of capital stock.

[9] R&D and outward FDI data is available a least for some of the group one countries, further analysis is not undertaken because the aim is to compare group one with group two.

[10] An alternative way of estimating domestic knowledge v_{it}^{D} could be based on the Kalman filter to infer the "quality" of the systems of innovation.

[11] See Appendix 1 for country sample.

[12] See Appendix 2 for further details.

[13] We note that $E\left(p_{it}|p_{it-1},V_{it},\theta_{t}\right)$ does not require that future exogenous variables be uncorrelated with disturbances: $E\left(V_{it}\omega_{is}\right)\neq 0$ *for all* $s < t$ so that a feedback mechanism is allowed from p_{it} to V_{it+1}. However, the explanatory variable $V_{it} \equiv \left(v_{it}^{M},v_{it}^{FDI},v_{it}^{D}\right)$ must be predetermined by at least one period: $E\left(V_{it}\omega_{is}\right)=0$ *for all* $s > t$)

[14] Although the Sargan test is satisfied, we note that one of its requirements is that the error terms must be homoscedastic whereas in our case they are heteroscedastic implying that the extent to which the test can confirm the validity of instruments is limited.

[15] See Table 2 in Appendix 2 for results of the two groups of developing countries.

[16] A similar method is used by Bernanke and Gurkaynak (2001) where $K_{1980} = I_{1981}/(g+d)$ where g is the growth rate of output and d the rate of depreciation.

[17] This computation assumes that perfect competition in factor markets and more closely reflects industrialised countries. It nevertheless, gives a reasonable indication of minimum acceptable rate of return to capital; the aim of the analysis is not to demonstrate the limitation of this assumption of the linear model. Rather the aim is to determine the implications of excluding a variable that represents domestic innovation (the variable is generally left out of estimation specifications for developing countries).

[18] In their empirical study, Lichtenberg and van Pottelsberghe de la Potterie (1996) found that inward FDI flows do not constitute a significant channel of technology transfer. While their study concerns industrialised countries there is reason to believe that the results would not hold for developing countries.

References

Abramovitz M. (1986), 'Catching Up, Forging Ahead, and Falling Behind,' in *Journal of Economic History*, 46(2): 385-406.

Arellano M. (1989), 'A Note on the Anderson-Hsiao Estimator for Panel Data,' *Economics Letters*, 31:337-341

Arellano M. and Bond S. (1991) 'Some Tests of Specification for Panel Data: Monte Carlo Evidence and an Application to Employment Equations,' *Review of Economics Studies*, 58: 277-297

Barro R. and Lee J.W. (2000), 'International Data on Educational Attainment: Updates and Implications,' *NBER Working Paper*, no. 7911.

Benhabib J. and Spiegel M. (1994), 'The Role of Human Capital in Economic Development: Evidence for Aggregate Cross-country Data,' *Journal of Monetary Economics*, 34(2): 143-173.

Benhabib J. and Spiegel M. (2002), 'Human Capital and Technology Diffusion,' *FRBSF Working Paper*, no. 2003-02.

Bernanke B. S. and Gurkaynak R.S. (2001), 'Is Growth Exogenous? Taking Mankiw, Romer, and Weil seriously,' in Bernanke B. S. and Rogoff K. (eds.), *NBER Macroeconomics Annual 2001*, Cambridge, MA: MIT Press.

Bloom D., Canning D., and Sevilla J. (2004), 'The Effect of Health on Economic Growth: A Production Function Approach,' in *World Development*, 32(1): 1-13.

Coe D., Helpman E., and Hoffmaister A. (1997), 'North-South Spillovers,' in *The Economic Journal*, 107(440): 134-149.

Cohen W. and Levinthal D. (1989), 'Innovation and Learning: The two faces of R&D,' in *The Economic Journal*, 99(397): 569-596.

Cohen W. and Levinthal D. (1990), 'Absorptive Capacity: A new perspective on Learning and Innovation,' in *Administrative Science Quarterly*, 35 (1): 128-152.

Devarajan S., Easterly W., and Pack H. (2001), 'Is investment in Africa too low or too high? Macro and micro evidence,' *Policy Research Working Paper – 2519*, Washington D.C.: World Bank.

Eaton J. and Kortum S. (1996), 'Trade in Ideas: Patenting and Productivity in the OECD,' in *Journal of International Economics*, 40 (3/4): 251-278.

Easterly W. and Levine R. (2002), 'It's not factor accumulation: Stylized facts and growth models,' *Working Papers-Central Bank of Chile*, no. 164.

Fagerberg J. and Verpagen B. (2004), 'Innovation, Growth and Economic Development: Why some countries succeed and other don't,' *Working Paper* 02/04, University of Oslo.

Freeman C. (1987) Technology Policy and Economic Performance: Lessons from Japan, London: Frances Printer Publishers.

Hausman J. and Taylor W. (1981), 'Panel Data and Unobservable Individual Effects,' *Econometrica*, 49: 1377-1398.

Kaldor N. (1957), 'A Model of Economic Growth,' in *The Economic Journal*, 67(December): 591-624.

Kaldor N. (1961), 'Capital Accumulation and Economic Growth,' in Lutz F. A. and Hague D. C. (eds.), *The Theory of Capital*, New York: St. Martin's Press.

Keller W. (2001), 'International Technology Diffusion,' *NBER Working Paper*, no. 8573.

Klenow P. and Rodriguez-Clare A. (1997), 'The Neoclassical Revival in Growth Economics: Has it come too far?,' in Bernanke B. and Rotemberg J. (eds.), *NBER Macroeconomics Annual*, Cambridge, MA: MIT Press.

Klenow P. and Rodriguez-Clare A. (2004), 'Externalities and Growth,' *NBER Working Paper*, no.11009.

Lall S. (2000), 'Industrial Success and failure in a Globalising World,' *QEH Working Paper Series*, QEHWPS 46, University of Oxford.

Lichtenberg F. and van Pottelsberghe de la Potterie B. (1996), 'International R&D Spillovers: A Re-examination,' *NBER Working Paper*, no. 5668.

Mankiw D. G., Romer D., and Weil D. N. (1992), 'A Contribution to the Empirics of Economic Growth,' in *Quarterly Journal of Economics*, 107(2): 407-437.

Mayer J. (2001), 'Technology Diffusion, Human Capital and Economic Growth in Developing Countries,' *UNCTAD Discussion Paper*, no. 154.

Mincer J. (1974) *Schooling, Earnings, and Experience*, New York: Columbia University Press

Mohnen P. (2001), 'International R&D Spillovers and Economic Growth,' in Matti Pohjola (ed.), *Information Technology, Productivity and Economic Growth: International Evidence*, UNU/WIDER in development economics, Oxford: Oxford University Press.

Nelson R. and Winter S. (1982), *An Evolutionary Theory of Economic Change*, Cambridge, MA: Harvard University Press.

Romer P. (1990), 'Endogenous Technological Change,' *Journal of Political Economy*,98(5)-Part II: S71-S102.

Rosenberg N. (1983) *Inside The Black Box: Technology And Economics*, Cambridge, UK: Cambridge University Press

Solow R. (1956), 'A Contribution to the Theory of Economic Growth,' *Quarterly Journal of Economics,* 70:65-94.

Solow R. (1957) 'Technical Change and the Aggregate Production Function,' *Review of Economics and Statistics,* 39:312-320.

Temple J. (1999), 'The New Growth Evidence,' *Journal of Economic Literature,* 37(1): 112-156.

UNCTAD (2004), *FDI Online Database,* Available at: http://stats.unctad. org/FDI//

Xu B. (2000), 'Multinational Enterprises, Technology Diffusion, and Host Country Productivity Growth,' *Journal of Development Economics,* 62(2): 477-493.

African Journal of Science, Technology, Innovation and Development
Vol. 1, No. 1, 2009
pp. 167-189

Building Systems of Innovation in an African Setting: The Cluster Initiative Development Approach

Bitrina D. Diyamett*

Abstract

Under the current competitive environment the word "innovation" has become catchy and center of attention for both researchers and policy makers. Much of the academic discourse and policy targets are on how to spur innovative activities among national firms. Although the debate on factors facilitating innovative activities is far from being settled, it has concretely been established that innovation is systemic and context specific and therefore its facilitation requires multidimensional and context specific approaches. This paper proposes that innovation in the African context can best be built through cluster initiatives. In building the argument, the paper brings together literature on systems of innovation and those on clusters and argues that although the two concepts are means to the same end, they have taken rather parallel roots. It further argues that while the cluster concept is practical, the concept of systems of innovation is basically an *ex-post* rather than *ex-ante* concept, and therefore cannot easily be applied in building systems of innovation, especially in the African context. The Tanzanian Cluster Initiatives Project that has started showing positive results is used to back the theoretical arguments made in the paper.

Keywords: Systems of Innovation, Cluster Initiative, Tanzania, Africa.
JEL Classification: O33, O55, R11

1. Introduction

As the process of globalization gathers momentum and competition among nations become fierce, the issues surrounding innovation are

* Senior Research Officer at the Tanzania Commission for Science and Technology, and a National Coordinator for the African Technology Policy Studies Network, Tanzanian Chapter. Emails: bdiyamett@costech.or.tz; atps@daima.co.tz

receiving ever-greater attention from both researchers and policy makers[1]. This is from the fact that, innovation is considered as one of the most important competitive edges of nations. When it comes to competition, the issue is no longer price of goods and services, but quality. Much of the debate and policy targets are on how to spur innovative activities among national firms. Although the debate on factors facilitating innovative activities is far from being settled, it has concretely been established that innovation is systemic and context specific, and therefore its facilitation requires multidimensional and context specific approaches. This paper is a modest attempt to propose how innovation systems can be built in an African context. It does this by bringing together cluster and innovation theories. In so doing, the paper revisits some theories on innovation and on clusters. The major objective is to attempt to bring the two together for better insights on how best to facilitate technological innovations in an African context. The paper proceeds to paint a rough picture of innovation systems of poor developing countries such as those in Africa; and then suggest cluster initiative approach as one of the best vehicles for initiating and nurturing innovative activities in Africa. This rather theoretical suggestion is backed by lessons learned as a result of pilot cluster initiatives project currently being implemented in Tanzania. Although the cluster initiatives are still at a very early stage, and not much in terms of economic gain is evident, early innovation indicators such as trust and linkage formation are already emerging. It is important to state outright that the major objective of this paper is not to propose concrete policy options for strengthening/building systems of innovation in an African context, but rather to instigate discourse on this very pertinent issue, especially the role of clusters in innovation systems development.

The rest of the paper is organized as follows: The next section presents a theoretical framework consisting of two strands of literature: those on innovation studies, and those on cluster studies. Section three is a brief recapitulation of an African setting as far as innovation system is concerned. The argument for cluster initiatives approach for nurturing innovation, especially for Africa is further emphasized here. Section four presents some lessons learned through the pilot project on cluster initiative in Tanzania; this section also introduces the concept of "science parks" as a vehicle for introducing high tech based cluster initiatives. Finally section five draws some concluding remarks and makes some recommendations.

2. Theoretical Framework

The theoretical framework proposed consists of two strands of literature: those on systems of innovation, and those on clusters. In concluding the section the two strands of literature are pulled together, to focus on how cluster initiatives approach is used to develop innovation system, especially for poor developing countries such as those in Africa.

2.1 The Concept of Technological Innovation

The essence of innovation is novelty. It is both the process of introducing something new and useful and the new thing itself. It is a concept of very general application and there is wide range of approaches to conceptualizing innovation in the scholarly literature (Fagerberg *et al.*, 2004). Innovation has also been studied in different contexts, including in relation to technology, social system, economic development, and policy construction. One can talk of innovation in education system, constitutional innovation, etc. This paper is mainly concerned with innovation in economic context, defined as successful creation, development, and marketing of new goods or successful application of new techniques or ways of working that improve the effectiveness of an individual and organization (Archibugi *et al.*, 1994). Inbuilt in this definition are four major types of innovation: product innovation, process innovations, market innovation, and organizational innovation. Of the four types, product and process innovation are the most important. Non- technological innovation such as market and organizational innovation support product and process innovation (Teese, 1998; Lam, 2004).

Innovation can also be of differing degrees. It can be radical (completely new, but within the same techno-economic paradigm), and fundamental (completely new and cause a paradigm shift). An example of a fundamental innovation is the discovery of a transistor, which is the basis of the emergence of the IT industry. Innovation can also be incremental (small modifications of existing technology). Incremental innovations are achieved through learning by doing and learning by using (Rosenberg, 1982). According to Rosenberg (1982), the cumulative effects of these small changes can be as large and bring productivity growth that major innovations are capable of generating. It is a source of technological innovation that is not usually explicitly recognized as a component of the R&D process, which overlaps with the development and receives no direct expenditure as it does not involve major investment, which may be the reason why it is normally ignored.

2.2 Drivers of Innovation and the Role of Clusters

This section will briefly discuss models of innovation and the concept of clusters independently. It then ventures into a rather more difficult task of bringing the two concepts together, highlighting the pros and cons of each; and finally concludes that, clusters, through cluster initiatives (CIs), is the best approach to build systems of innovation in an African setting.

Models of Innovation

Models or motive forces for innovation is a major area of concern for policy makers, and consequently has been, and still is, a major area of inquiry in academic discourse. The debate started with a linear model of innovation where innovations were thought to be triggered as a result of basic research in science resulting in a widespread marketing of new products and widespread use of new processes, and by implication economic growth (Coombs *et al.*, 1987; Rosenberg, 1982). Science was thought to be an endless frontier; and all governments needed were a lot of scientists with well equipped laboratories. This belief seems to have been instigated by the success of the Second World War military science, especially the Manhattan Project to develop the atomic bomb – the most destructive weapon ever made by a group of scientists and engineers, the first time in history. The Manhattan Project was a demonstration that scientific projects could be large, hierarchically organized and yet achieve pre-determined goals (Hallonsten, 2009). Later on, and as a consequence of the post war science, in 1945 Vannevar Bush - the then Director of the United States Federal Office of Scientific Research and Development - published a report: *Science, the Endless Frontier,* which laid the basic principles of the partnership between science and the state, and was incorporated in the American science policy as the document setting the course for the United States' development into scientific super power (Guston, 2000; Hallonsten, 2009).

However, as things unfolded in the North, especially after the military technology had trickled down to the civilian science and technology, the notion of linear model of the *push type* was criticized as being inadequate in explaining what was really happening, and the linear model of the pull type, where the market was considered to be decisive in making innovation happen, was introduced. The demand pull model emphasizes the dynamics of market demand as the catalyst for techno-logical change - the changing market potential guides innovative activities to most profitable areas (Schmookler, 1966). This model was again later on criticized as inadequate; leading to what was referred to as *interactive* model of innovation. It was argued that innovation happens as a result of complex interaction between the 'supply' (R&D labs, scientific and technical institutions) and the 'demand' (potential and actual users, and marketing organizations) (Rothwell and Zegveld, 1985; Lovio, 1985). This must have led to the birth of a more recent concept of National Systems of Innovation (NSI) which has become popular, and almost exclusively being used to analyze different innovation systems. Lundvall, a pioneer of the concept, defines *NSI as that system which is constituted by elements and relationships between those elements, which interact in the production, diffusion, and use of new and economically useful knowledge* (Lundvall, 1992: 2). According to many proponents of NSI, the elements are largely of two types: organizations and institutions. Organizations are formal structures, where innovation related activities take place. Some important organizations in the SI are companies, which can be producers, suppliers, customers, and competitors in relation to other companies; universities; R&D organizations; venture capital organizations and public innovation policy agencies. Institutions are set of common habits, routines, established practices, rules or laws that regulate the relations and interaction between groups and organizations. The examples in the SI are for instance, incentive structures that influence innovative activities, patent laws and norms influencing the relations between universities and firms (Edquist, 1997). The concept of NSI, as observed by some scholars, cannot be generalized universally. It emphasizes the specificity of innovative activities to a particular socio-economic and social-cultural context. Although the necessary types of organizations and institutions largely remain the same, their relative importance and the way they interact very much depend on the existing socio economic environment (Nelson, 1993; Lundvall, 1992; Gu, 1999; Edquist, 1997). Edquist (1997) also talk of the functions to be performed by the actors of the NSI. These are:

1. Provision of R&D and creating new knowledge.
2. Competence building (provision of education and training, creation of human capital, production and reproduction of skills, individual learning) in the labor force to be used in innovation and R&D activities.
3. Formation of new product markets

4. Articulation of quality requirements emanating from the demand side with regard to new products.

5. Creating and changing organizations needed for the development of new fields of innovation, e.g. enhancing entrepreneurship to create new firms and entrepreneurship to diversify existing firms, creating new research organizations, policy agencies, etc.

6. Networking through markets and other mechanisms, including interactive learning between different organizations.

7. Provision (creation, change, abolition) of institutions, e.g. IPR laws, tax laws, environment and safety regulations, R&D investment routines, etc. that influence innovating organizations and innovation process by providing incentives or removing obstacles to innovation.

8. Incubating activities, e.g. providing access to facilities, administrative support, etc. for new innovating efforts.

9. Financing of innovation process and other activities that can facilitate commercialization of knowledge and its adoption.

10. Provision of consultancy services of relevance for innovation process, e.g. technology transfer, commercial information and legal advice.

The above ten functions can selectively be applied in building systems of innovation depending on the context and extent to which the system has been developed. The functions are especially important in identifying gaps in the required organizations and institutions for a certain system of innovation to perform well. We will return to this point when discussing how cluster initiatives are used to build systems of innovation.

As earlier explained, the above developments in the theories and policies on technological innovation to a large extent were based on the empirical observations of the changes in practice; in the now developed countries – *academic discourse and policies chased practice* while, as we shall see shortly, in most African countries, such as Tanzania, policies are chasing theories developed from the empirical observation based on very different contexts. Worse still, the theories claimed to have become redundant in the context in which they originated. A good example is the over emphasis on scientific research and training of more scientists as a way of fixing the African technological problem. This practice seems to be rooted in the Bush report alluded to above, which has already been overtaken by events. This is ironic because it is well known that the greatest technological problem in Africa is the link between science and the technology, and not the inadequacy in scientific research.

The Concept of Clusters

Clusters are groups of firms engaged in similar or related economic activities in a national economy. In most cases they are defined by two important attributes, namely spatial agglomeration and sectoral dimension. According to Rosenfeld (1997), cited in Le Veen (1998), an industry cluster is a geographically bounded concentration of similar, related or complementary businesses, with active channels for business transactions, communications, and that are faced with common opportunities and threats. According to Porter (1990), clusters are basically of two types: horizontal and vertical. Vertical clusters are made up of firms that are linked through buyer-seller relationships. Horizontal clusters include firms that share common market for the end products, use common technology or labor force skills or require similar natural resources; they are basically competitors. However, in most cases clusters are made up of both horizontal and vertical relationships. In short they are firms, their suppliers, users and competitors co-located geographically. Clusters have, in most cases, been associated with dynamism, innovation and competitiveness, basically because of collective efficiency, linkages and externalities that enable enterprises to overcome many constraints in the area of capital, skills, technology, markets, etc. Three advantages of firms locating in clusters can be identified in the literature. The first advantage is the intensity of a labor pool due to geographical concentration of firms in the same industry or in closely related ones. The second advantage is the availability of related materials and other inputs at lower costs. These inputs include tangibles like raw materials and supplies, and intangibles like consultation and collaboration. The third advantage is the intensity of knowledge exchange that can lead to knowledge spillovers between nearby firms and institutions in the cluster (David *et al.*, 1990; Krugman, 1991; Kelly *et al.*, 1999; Oyeyinka, *et al.*, 2007).

Whereas all clusters, as indicated above, may have properties that serve to speed innovation, some can be observed to be particularly prone to innovation (Anderson *et al.*, 2004). This has led to the coining of the concept of *innovative clusters*. An innovative cluster innovates in the broadest sense of the concept as earlier discussed. That is, innovation can emanate from improvements in the way that actors organize themselves; new products are developed, produced, commercialized and distributed.

An innovative cluster is, in principle, evolving constantly, learning from experiences and able to adjust to changing circumstances. It is likely to be well positioned to explore new opportunities beyond its present boundaries and at the same time, combine flexibility with inner strengths, stability and sense of direction to achieve the following:

- Traditional boundaries to knowledge generation and diffusion are continuously changed by establishing linkage to wider and alternative sets of knowledge inputs.
- Products and markets are re-conceptualized.
- Mechanisms for seed-funding, risk-taking and entrepreneurship are upgraded.
- Old institutions and organizations are transformed through learning as well as unlearning of earlier habits and practices (Anderson et al., 2004).

The above transformations can be aided through cluster initiative development, the subject we will shortly return to.

(a) Clusters and Systems of Innovation

The concept of clusters and systems of innovation are somehow related; and perhaps are means to the same end, which is competitiveness. However scholars of the two concepts have stressed two different things in the same value chain. While scholars of clusters have stressed competitiveness directly and rarely mention innovation explicitly, scholars of innovation systems have stressed innovation as means to achieve competitiveness; and therefore the major focus has always been innovation and how best to nurture it. In terms of effectiveness in facilitating competitiveness, the two concepts have pros and cons. The cluster concept is much narrower than the innovation system. It explains, and can be used to facilitate innovative activities at a micro level, while innovation system takes into account circumstances and institutions at a more macro level. However, while the concept of systems of innovation is so rich in explaining circumstances and conditions under which innovation can occur so that nothing is left out in trying to put a jigsaw-puzzle of innovation together, the concept is so loose to the extent that its proponents have failed to pin it down in terms of how to use it to facilitate innovative activities. Some of these weaknesses have been acknowledged by its pioneers. For instance Lundvaal et al. (2002: 226) acknowledge that the concept has not yet been applied to system building - it has mainly been used as an ex-post rather than ex-ante concept. To the contrary, the concept of clusters has been, as we will shortly see, operationalized through cluster initiatives; and experience has shown that it works.

In concluding this section, it suffices to say that *innovative cluster* can be equated to the well performing *systems of innovation*, especially at the micro and local levels. We have also mentioned that, while there are not yet any concrete practical ways to facilitate innovative activities using the concept of systems of innovation, there are practical ways that have been proposed to achieve innovation through the concept of cluster initiatives. The following section discusses issues surrounding this concept. Major emerging issues will be used to explain early success stories in the Tanzanian cluster initiative project.

2.3 Development of Cluster Initiatives or Building Systems of Innovation

As earlier alluded to, many clusters evolve spontaneously and take shape gradually over an extended period of time. The process falls into 4 main phases: (i) creating trust; (ii) forming linkages; (iii) vision or strategic direction; and (iv) undertaking actions (Anderson *et al.*, 2004). However, recently, there has been a wide spread practice in externally intervening in this otherwise natural process. This conscious external intervention is popularly known as *cluster initiative*. The interventions are geared towards speeding up the process of a development of a cluster, or preventing it from stagnation and decay. Generally, the process is intended to instill dynamism in a cluster and make it more innovative and competitive. Various actors who relate to each other in different ways carry out the intervention. Governments and other public authorities are known to be responsible for most cluster initiatives, although there are marked geographical variations. For example in the U.S the private initiatives are more common (Solvell *et al.*, 2003, cited in Anderson *et al.*, 2004). However, even if the initiative is started by a non-government entity, to a large extent the initiative will still depend on the government in terms of provision of funds and other incentive structures. The government is therefore a very important agent in cluster initiatives.

This paper argues that, by analogy, cluster initiatives development can also be termed as building systems of innovation at the micro level. The cluster initiative, or system of innovation development can start at any given phase of the naturally developing cluster; and in the literature there is no one fit approach to trigger the process. There are however three major alternative cluster initiative processes: (i) the engineered; (ii) the organic; and (iii) the re-engineered. Each goes through the same general phases mentioned above, but with different entry points according to the stage at which the cluster is at the time of intervention. It is very important to recognize the phase at which a certain cluster is at the time of intervention as this gives a rough idea - with the assistance of the ten functions of systems of innovation earlier mentioned - of what activities to start with, and what new organizations and institutions need to be put in place. Below we discuss different cluster initiative processes and their entry points.

The Engineered

This is generally a top-down cluster initiative approach. It is typical of those clusters that are still at the early stage of development. In Europe, policy makers lead and in North America, it is the private individuals who are important. In Africa who leads or rather who can lead? University? NGO's? Development Partners? The answer to this question is not yet apparent at this point in time as cluster initiatives is a very new concept in Africa. However, as earlier alluded to, and as will shortly be evident through Tanzanian cluster initiatives pilot project, government is a very important actor. While cluster initiatives can be started by any of the above mentioned actors, for sustained interventions and growth the government has to come in, in one way or the other.

The initial catalyst of an engineered cluster process could be a given investment opportunity, a dynamic leader, or a regional/national economic crisis (poverty for Africa). The general steps are as follows:

- Form or develop existing social capital to anchor the cluster idea.
- Maintain or establish new mechanisms for building trust, formulate a vision and strategy and then undertake action.

The Organic

This is a bottom-up approach. These clusters initially display sponta-neous development towards the establishment of linkages and joint strategy. From this platform of continuous or re-curing instance of cooperatives, an innovative cluster appears. Intervention or cluster initiative is targeted at the tightening of networks and collaboration, introduction of supportive framework, acquisition of things such as new technology, and removal of rigidities.

The Re-engineered

The re-engineered cluster is when an existing cluster (engineered or organically developed) is viewed as having specific competitive significance or potential, but is hindered from progress for some critical reasons. Key linkages are broken or never formed, or there are other crucial delimiting factors within the cluster itself or its surrounding that are blocking its dynamism. For such reasons the process is re-started through corrective actions such as re-establishing, or establishing key linkages, dismantling or breaking of adverse rigidities; or through the communication of new vision and strategy. The whole picture of these processes with different entry points can be represented diagrammatically as shown in Figure 1 below. The figure is very useful in indicating entry points for external intervention for clusters at different stages of natural development.

Figure: 1: Cluster initiatives different entry points

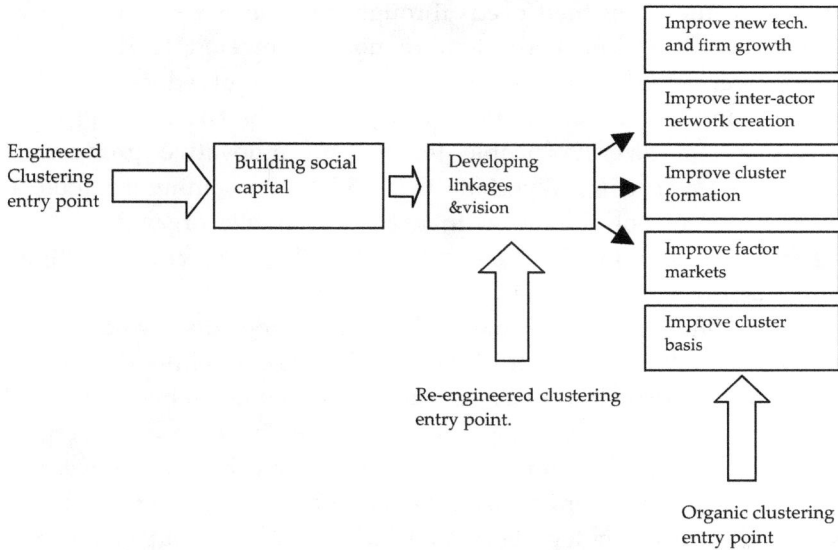

Source: Adapted from Anderson *et al.*, (2004).

As earlier discussed, innovation is context specific; therefore before we put forward our argument that cluster initiatives development is the most efficient way of building systems in Africa, we will discuss, albeit briefly, the African context within which the innovation system is placed.

3. System of Innovation in an African Context: The Reality

There are debates in the literature on whether or not anything like systems of innovation exists in the least developed countries. The debate seems to be in favor of our contention that as long as some form of innovation exists in the least developed countries, then some form of a system of innovation must exist (see for instance, Lundvall *et al.*, 2002). The real disadvantage of Africa in this is the lack of systematic studies on innovation in the context of Africa.

Having said the above, the real intention of this section is not to discuss innovation systems in an African setting, but rather, as mentioned above, to discuss the context in which innovation systems are placed. African context is characterized by two major system weaknesses that have a direct bearing on the most basic forces behind innovative activities, i.e. *technology push* and *demand-pull*. Briefly, the first weakness is that African economies are very fragile, the markets are small and underdeveloped. While to a large extent, consumers in more developed countries can express their needs through clear-cut market mechanisms, those in most African countries are not empowered to do so. This disempowerment largely manifests in low level of education of the populace and low per capita income, which are the two most important aspects in empowering people to demand innovative products. In addition, infrastructure, especially those related to communications, are weak, R&D are not conducted in firms, but public organizations and largely divorced from the productive sector (Wangwe *et al.*, 1998); and there is lack of venture capital organizations.

The second important weakness is a conspicuous disconnect between science and technology in most African countries. It is important to note that there are currently radical global developments taking place in the field of science and technology. There is ever increasing convergence between science and technology. In the sense that the line between the two has become extraordinarily thin - science is increasingly becoming technology and technology becoming science. Good examples are fields of biotechnology and nanotechnology. This is not a problem for the developed countries because science and technology are more or less intertwined - it is where the convergence between science and technology is currently happening. The situation in Africa is different. While the scientific community in Africa to a large extent is moving ahead with the rest of the scientific community globally[2], technology in Africa has to a large extent remained local, and at a much lower level and divorced from science. But to avoid long term or permanent marginalization under increasing globalization, Africa must endeavor to build organic relationship between science and local technology.

Apart from the above mentioned weaknesses, the African environment is different in the sense that it consist of many diverse subsystems such as an agricultural sector (food and cash crops) and an industrial sector that largely consists of manufacturing and mining. Within the industrial sector, there are also large and small-scale enterprises, which are both formal and informal. In short African innovation systems consist of many different sub-systems, with different actors, problems and challenges, hence requiring a different set of policies to facilitate innovative activities. For these reasons, it is much easier to facilitate and manage innovation at a more micro level depending on the above different contexts. For this reason, in an attempt to build systems of innovation in an African setting, we recognize the special relevance and importance of cluster initiatives, especially the innovation enhancing aspect of clustering[3]. Clusters are especially important vehicles for the African setting for the following reasons: First, is the already discussed fact that the concept of systems of innovation is a bit abstract and vague, and difficult to operationalize, especially in an African context, where there are diverse sub-systems; second, projects draw attention of the governments and donors for some practical actions. If it goes in the name of a cluster initiative project, innovation system building will have a better chance of succeeding than general innovation policies that are normally not backed by systematic empirical studies of the African situation. Moreover, most policies in Africa lack implementation strategies. A cluster initiative approach will enable Africa to move beyond the meaning of policy as simply *nicely packaged blue prints* to *a course of action,* where lessons that emerge as clusters evolve are continuously being incorporated into the cluster policies. The ten functions of systems of innovation earlier alluded to can help in putting in place missing institutions and organizations as clusters grow towards maturity. Lastly, the initiative can be started by any body as a demonstration to attract the attention of those who matter, just as it was done by the College of Engineering and Technology of the University of Dar Es Salaam but managed to attract the attention of the government and development partners.

4. Some Lessons from a Pilot Project on Cluster Initiatives in Tanzania

In this section we make reference to the ongoing cluster initiatives project in Tanzania. The section also proposes a methodology of initiating high tech clusters.

4.1 The Current Project

Tanzania embarked on the cluster initiatives project in early 2006 through the generous support of SIDA. During the 18 months of the pilot project, 8 clusters were selected for the cluster initiatives. These clusters are: Metalworking and Engineering Cluster, Mushroom Cluster, Vegetable Seed Cluster, Seaweed Cluster, Tourism and Cultural Heritage Cluster, Nutriceuticals Cluster, Sisal Cluster, and Vegetable and Fruits Cluster. The project started with the training of the cluster facilitators on the cluster initiative concept and the identification of "low hanging fruits"; that is, identification of the immediate needs with lowest cost possible by the cluster facilitators. The progress of the 8 clusters was monitored over the 18 months life of the project. At the end of the 18 months, the project was evaluated; and 4 clusters were chosen for upgrading. This paper draws lessons from this final evaluation to argue the case for emerging innovation systems[4]. Although it is too early to draw any conclusion, the project seems to have put the participating actors on the right path towards stronger innovation systems with a potential to develop much stronger in future. As earlier alluded to, clusters evolve and take shape gradually over an extended period of time.

Important lessons learnt during the project evaluation include the increased visibility of actors, trust and linkage formation, collective efficiency and product diversification. It is important to note here that, as already discussed; these factors are at the core of the innovation capability building. The following paragraphs explain each of these:

Visibility

The project has been instrumental in popularizing the cluster concept, and in increasing the visibility of firms and farms in the project, and has succeeded in drawing the attention of government officials and donors. As a result, some clusters have already received some assistance from government agencies with some positive impacts. A good case is the mushroom cluster where the cluster accessed a building to act as a mushroom collection center from the local government. This cluster also accessed funds for training and production of the mushroom growers' manual from the Small and Medium Enterprises (SME) Competitiveness facility. The Morogoro Food processing cluster acquired a premise for a joint market outlet from the local government in Morogoro region. Even though, the assistance has not been much, and tangible economic gains are less visible at this point in time, social and political capital cultivated so far promise to be of great value in the future.

Trust and Linkage formation

Trust and linkage among firms and farms in clusters are the greatest capital for innovation and prosperity. It is encouraging to see that the project has achieved a noticeable level of trust and linkages among cluster members. In the words of one cluster member of the mushroom cluster during one of the focus group discussions:

> Not only has our business relations improved, but also our social life. It was not common for me to talk to some of the cluster members, not even to exchange greetings, but now we are buddies, and I feel free to confront them with whatsoever problem I have, whether social or business.

This seems to be a sentiment shared by most of the cluster members in all the 8 clusters. For clusters such as mushroom and seaweeds farming, where the business is so much knowledge intensive that a small mistake can ruin the whole business, clusters have most visible impacts on the improvement of the products' quality because of enhanced sharing of information and learning from each other. Through linkage and learning, the banana farmers also gained more income by discovering that banana leaves are used in mushroom farming. The Morogoro Food processing cluster has now a joint marketing outlet for their products, which is a better marketing strategy, especially for small producers.

In the case of the Morogoro metal cluster, there is now an improved cooperation between the cluster members, especially between the small and larger firms. In the words of one cluster member, before the project, it was unthinkable for small entrepreneurs to approach the larger firms for help, but now it is much easier. Generally the incidence of cooperation such as joint purchases of inputs and job sharing in cases where orders go beyond individual capacity has increased among cluster members. There has also been some technology transfer, in terms of training as a result of the project, and the sharing of information among the cluster members.

The cluster initiatives project has also enhanced linkage with the higher learning institutions for some clusters, even though there is mixed feelings in this. While some clusters are happy with the assistance they receive from the university, others complain that equipment from the university are more expensive and not necessarily superior to same equipment from elsewhere in the country. This complaint has also been observed by a study on public perception of the roles of universities in socioeconomic development in Tanzania (Mwamila and Diyamett, 2009). This point to the importance of the relationship between science and technology earlier alluded to.

Collective efficiency

Another important attribute of a cluster is collective efficiency, which reduces the cost of doing business, and also leads to innovative activities especially in subcontracting and collective branding. This is already happening with some clusters in Tanzania. The existing collective efficiency is in the form of a joint market outlet (Morogoro Vegetable and Fruits, and Mushroom); joint purchase of raw materials (Morogoro Metal Work); job sharing and subcontracting (Morogoro Metal Work); joint exhibition (Morogoro Metal Work, Mushroom and Morogoro Vegetable and Fruits) and collective branding, which is in the process (Morogoro Vegetable and Fruits cluster).

Product Diversification and Quality Improvement

The cluster initiative project has led to an increased product diversification and increased value chain in some clusters. Good examples are mushroom and seaweed clusters. For mushroom cluster for instance, new mushroom products such as cakes and other snacks have been developed by the clusters. This is the result of training, and information sharing among cluster members. For the seaweed cluster, the project has led to improved quality of seaweed, which stems from the introduction of appropriate drying techniques: Before the cluster initiative, farmers used to dry seaweeds by spreading them on the ground to be sun-dried, and in the process they collected dirt. Through the cluster initiative project farmers were introduced to new drying techniques whereby seaweed contamination with dirt is avoided. This has the potential of raising prices. The project has also introduced deep water farming where floating raft techniques are used. This has contributed to product quality improvement. The cluster is also moving towards high value addition activities. Traditionally, in Tanzania, seaweeds are farmed, harvested, dried and sold to traders ready for export. One of the positive achievements of the cluster is the ongoing efforts to add value by diversifying its local use. Though still work-in-progress, the proposed products include seaweed herbal soap products and seaweed snacks.

4.2 Initiating High tech Clusters

We earlier on argued that the Tanzanian systems of innovation, among others, consist of low tech sub-systems where innovation is incremental and achieved through learning by doing, rather than from R&D per se. This, to a large extent, is the case for all clusters in the above cluster initiatives projects. Although the cluster initiative project is being coordinated from the College of Engineering and Technology (CoET), University of Dar es Salaam, none of the clusters is a result of a commercialization of the research output from the University. However, there is the possibility for the university research output to come in as clusters move up the value chain, especially for the mushroom and seaweed clusters where there is clear indication of these clusters moving up the value chain. Otherwise the practice perpetuates the traditional de-link between science and technology that exists in most African countries. This paper argues that there is also a need to venture into cluster initiatives that will endeavor into the commercialization of new ideas emanating from the scientific research carried out by universities, especially for the cluster initiatives initiated and coordinated by a university such as this one. There are two major reasons for this: First, this is important if Africa is to be an equal player in the ongoing process of globalization. Second, commercialization of scientific research is important for closing the gap that exists between science and local technology. The gap does not only have serious implications for the country's socioeconomic development, but can also actually hamper scientific progress in the country. As earlier discussed, science and technology are interdependent, and science can only grow if it is put into practical application. From time in memorial, science and technology (S&T) have been influencing and benefited from each other. This is the reason S and T have always appeared together. In the early days, technology was a leader and science a follower. An example is the steam engine that was developed without any known scientific principle. On the other hand, the laws (science) of thermodynamics were discovered as a result of learning the working of the steam engine. Of course later on science became the leader and technology very much benefited from science. For this reason, if the organic relationship between science and local technology in Africa is not forged, to a large extent the direction of growth of science in Africa can be influenced by what is practical in developed countries. This is based on the fact that, as earlier alluded to, scientists in Africa collaborate more with scientists in developed countries than they do with peers within Africa; and we have already argued that science and technology are more or less intertwined in developed countries.

Having said the above, the intention of the paper is not to suggest concrete modalities of putting the cutting edge science in Africa into practical application, but rather to point out that this is an important research and policy gap in the Tanzanian systems of innovation, and to stimulate debate on the issue. This notwithstanding, the paper will attempt to point out two major generalities on the issue. First is on the approach and second is on priority areas.

One of the best approaches in creating a linkage between cutting age science and technology is through the concept of science parks. This is one form of a cluster initiative, which normally involves high technology companies. A Science Park consists of a space and buildings, normally close to a center of knowledge generation, such as the university. The structure, role, and positioning of a science park in a country's science, technology and innovation system very much depends on the social and economic environments of that particular country. But in most cases a science park will consist of university and other public centers' research groups; R&D units of companies; new spin off companies from universities, and new start-up companies from the existing business. All these in a space provided with top facilities to develop scientific and business activities. The major objectives of science parks include:

- Promoting R&D by the university in partnership with industry
- Assisting in the growth of new ventures
- Providing an environment where knowledge-based enterprises can develop in close interaction with a center of knowledge for their mutual benefit (UNESCO, 2007).

In terms of priority areas of investment for the Tanzanian environment, pharmaceuticals, agro-processing, mining and energy sectors could be chosen. These sectors have been chosen because the country has a comparative advantage in terms of the availability of natural resources, and they could also have profound impact on poverty reduction and growth strategies. The starting point could be to put in place a center for research commercialization within the university, where the first activities of this center can be analyzing all the university research outputs with the objective of identifying those with commercial potential.

5. Some Concluding Remarks and Recommendations

This paper has argued that the cluster initiative approach is the best available means of building systems of innovation in an African setting. Clusters have, in most cases, been associated with dynamism, innovation and competitiveness, basically because of collective efficiency, linkages and externalities that enable enterprises to overcome many constraints in the area of capital, skills, technology and markets. These are precisely the attributes of a strong innovation system. However, while cluster theories have elsewhere been put in practice, and seem to be working, the concept of the system of innovation so far has only been used to explain and not to build systems of innovation - it is an *ex-post* rather than an *ex-ante* concept. This is, but one reason for proposing a cluster approach for building systems of innovation in Africa. Another important reason is that African systems consist of diverse sub-systems, making it more suitable for the use of micro approaches, where cluster initiatives fit better than policies for the development of system of innovation, which is normally applied at a more macro level.

However, although the project is being coordinated from the University, none of the clusters is the result of the commercialization of the university research. This practice perpetuates the traditional de-link between science and technology. But for Africa to avoid marginalization and join the global value chain in production and marketing of products, it has to start investing in the commercialization of its research outputs. Three important actors have a role to play here. These are universities, governments and the private sector. But universities, especially through spin-off companies, have the upper hand here. Universities have to conceive strategies to commercialize their potential research results. Universities, especially the faculties of science and engineering, in collaboration with the faculties of commerce and economics should join hands to commercialize their research results. This however is not to argue that universities should work in isolation, but rather to argue that universities need to be proactive and sensitize the governments and the private sector on the need and potentials of commercializing research results emanating from their labs. Universities can demonstrate the viability of commercializing research results by starting spin-off companies; and governments have a very important role of putting in place incentive structures, including provision of venture capital.

Finally, we make three major proposals: two on policy implications and one in the area of further research. On policy, there is a need to encourage cluster initiatives in two distinct areas: First is the low technology sub-systems, where innovation is largely incremental, and achieved through learning by doing. These clusters can be managed by many other different bodies in the country such as public research institutions, commissions or councils of science and technology, and other relevant ministries. Second is the high tech innovation sub-system where scientific research emanating from universities and other similar institutions of higher learning are commercialized. These have to be managed by the universities themselves. In this regard the concept of science parks that is increasingly being referred to in Africa must now start taking shape.

The final proposal is on areas for further research. In this regard we propose that academic discourse on innovation and Africa should think outside the current *systems of innovation* box that has been developed in the context of the mainly developed countries.

Notes

[1] Although this is much less obvious in the African context, especially the academic discourse, politicians are increasingly paying attention to issues on innovation, which is a sign that the role of innovation in development is increasingly being recognized even in the African context. Academic discourse, especially innovation policy researchers need to take this opportunity to complement increasing political recognition of the role of innovation.

[2] Most scientists claim to belong to the "international" scientific community (Gillard, 1994) and they collaborate extensively, where the motive for collaboration is internal to science, rather than policy influence. However, while collaboration among scientists within the developed countries is more common, African scientists' collaboration is more skewed towards collaboration with the developed world, and they compete among themselves for collaborators with the more developed world. They seem to be doing so more for access to funding than recognition (Wagner and Leydesdorf, 2004).

[3] This paper argues that for Africa, at this point in time where the continent is only building foundations for competitiveness, the major issue is to build and strengthen conditions for innovativeness.

[4] Much of the materials in this section are derived from an ATPS-Tanzania cluster project evaluation report by Diyamett and Komba (2007). We acknowledge the authors of this report, ATPS-Tanzania, and above all, SIDA who financially supported the evaluation process.

References

Archibugi D., Evangelista R., and Simonetti, R. (1994),'On the Definition and Measurement of Product and Process Innovation,' in Shionoya Y. and Perlman M. (eds.) (1994), *Innovation in Technology Industries and Institutions: Studies in Schumpeterian Perspectives,* Ann Arbor, MI, USA: University of Michigan Press.

Anderson T., Serger S.S., Sorvik J., and Hanson, W.E.(2004), *The Cluster Policies White Book,* Malmo, Sweden: IKED.

Coombs R., Saviotti P., and Walsh V. (1987), *Economics and Technical Change,* London: Macmillan Education Limited.

David P. and Rosenbloom J. (1990), 'Marshalian Factor Market Externalities and the Dynamics of Industrial Location,' In *Journal of Urban Economics,* 28 (3):349-370.

Diyamett B.D. and Komba A. (2007), Tanzania Cluster Initiatives Project: An Evaluation Report of the 8 Cluster Initiatives, Dar es Salaam: ATPS-Tanzania.

Edquist C. (1997), Systems of Innovation, Institutions and Organizations, London: Pinter Publishers.

Fagerberg J. (2004), 'Innovation: A Guide to the Literature,' in Fagerberg J., Mowery D. C., and Nelson R. R. (eds.), *The Oxford Handbook of Innovations*, Oxford: Oxford University Press.

Gaillard, J. (1994), 'The Behavior of Scientists and Scientific Communities,' in Solomon J., Sagasti R., and Jeanlet C., (eds.), *The Uncertain Quest: Science, Technology and Development*, Tokyo, Japan: UNU Press.

Gu S. (1999), 'Implication of National Innovation Systems for Developing Countries: Managing Change and Complexity in Economic Development,' *Discussion Paper Series*, UNU/INTECH, Maastricht, The Netherlands.

Guston, D. H. (2000), Between Politics and Science: Assuring the Integrity and Productivity of Research, Cambridge: Cambridge University Press.

Hallonsten, O. (2009), Small Science on Big Machines: Politics and Practices of Synchrotron Radiation Laboratories, PhD Thesis, Research Policy Institute, Lund University

Kelly, M. and Hageman A. (1999), 'Marshallian Externalities in Innovation,' in *Journal of Economic Growth*, 4(1): 39-54.

Krugman P. (1991), 'Increasing Returns and Economic geography,' in *Journal of Political Economy*, 99 (3): 483-499.

Lam, A. (2004), 'Organizational innovation,' *Working Paper No. 1*, School of Business and Management, Brunel University, UK.

Lovio R. (1985), 'Emerging Industries, Interactions between Productions, Technology and Markets in a Small open Economy', Technical Research Center of Finland, Research Notes 554, Espoo.

Le Veen J. (1998), 'Industry Cluster Literature Review,' *Urban and Rural Development*, 261 (March).

Lundvall B.-Å. (eds.) (1992), National Systems of Innovation: Towards a Theory of Innovation and Interactive Learning, London: Pinter Publishers.

Lundvall B.-Å., Johnson B., Andersen E., and Dalum B. (2002), 'National Systems of Production, Innovation and Competence Building,' in *Research Policy*, 31 (2): 213-231.

Mwamila, B.L. and Diyamett, B.D. (2009), 'Universities and Social Economic Development in Tanzania: Public Perceptions and Realities on the Ground,' in *Science and Public Policy*, 36(2):85-91.

Nelson R.R. 1993 (ed.), *National Systems of Innovation: A Comparative Analysis*, Oxford: Oxford University Press.

Oyeyinka B.O. and McCormick D. (eds.) (2007), *Industrial Clusters and Innovation Systems in Africa: Institutions, Markets and Policy*, Tokyo, Japan: United Nations University Press.

OECD (1997), Proposed Guidelines for Collecting and Interpreting Technological Innovations Data, Paris: OECD.

Porter, M. (1990), *The Competitive Advantage of Nations,* New York: Free Press.

Rosenberg N. (1982),*Inside the Black Box: Technology and Economics,* New York: Cambridge University Press.

Rothwell R. and Zegveld W. (1985), *Reindustrialisation and Technology,* Harlow: Longman.

Rosenfield S.A. (1997), 'Bridging Business Clusters into Mainstream of Economic Development', in *European Planning Studies,* 5(1): 3-23.

Schmookler J. (1966), *Invention and Economic Growth,* Cambridge: Cambridge University Press.

Teece D. J. (1998), 'Design issues of innovative Firms: Bureaucracy, Incentives and Industrial Structure,' in Chandler A. D., Hangstom P., and Solvell O. (eds.), *The Dynamic Firm,* Oxford: Oxford university press.

Solvell O., Lindqvist G., and Ketels C. (2003), *The Cluster Initiative Green Book,* Stockholm: The Competitiveness Institute (TCI) and Vinnova.

Wangwe S.M. and Diyamett B.D. (1998), 'Cooperation between R&D Institutions and Enterprises: The Case of URT,' in Atlas XI Bulletin, *Approaches to Science and Technology Cooperation and Capacity Building,* New York and Geneva: UNCTAD.

Wagner C. and Leydesdorf L. (2004), 'Self Organization and the Growth of International Collaboration in Science,' 33, ASCoR, University of Amsterdam.

UNESCO (2007), 'Asia-Pacific Regional Workshop on Science and Technology Park: Project Document,' Jakarta, Indonesia.

African Journal of Science, Technology, Innovation and Development
Vol. 1, No. 1, 2009
pp. 190-209

How Much Is Free? Quantifying Open Source Software Development in South Africa

Michael Gastrow* and Saahier Parker**

Abstract

In the context of increasing globalisation and the emergence of knowledge-based economies, the promotion of open source software (OSS) has received increased attention internationally. There exists a broad consensus that OSS has an important role to play in the development strategies of emerging economies, and in many countries, including South Africa, this has been manifested in policies supporting domestic open source software development. However, the current policy has been developed in the context of a paucity of reliable quantitative data describing open source software development in South Africa. This paper therefore uses national Research and Development Survey data to provide a quantitative overview of open source software development in South Africa, with a view to informing future research directions in this area. The findings are supported by interviews with product development executives from four of South Africa's major software development firms. Key findings include a recent increase in expenditure on open source software development, and the identification of private sector clusters of activity in Gauteng and the Western Cape. Identified areas for potential further research include regional-level business-sector surveys, a focus on micro-level OSS developers, increased monitoring and evaluation of government OSS policies, and closer examination of the interaction between non-profit OSS advocates and government institutions.

Key Words: Open source software, software development, quantifying open source software, South Africa.

JEL Classification: O33, O34, O55

* Education, Science and Skills Development Research Unit, Human Sciences Research Council, South Africa. Email: mgastrow@hsrc.ac.za

** Centre for Science, Technology and Innovation Indicators, Human Sciences Research Council, South Africa. Email: sparker@hsrc.ac.za

1. Introduction

Open Source Software (OSS) is in essence a form of software where the source code of the software is available to users, and it is therefore possible for users to modify the software to suit their needs. In many cases this software is also free, or has weaker intellectual property rights (IPRs) attached in comparison to traditional proprietary software (for a more detailed definition see Krogh and von Hippel, 2003).

OSS has already had an enormous impact on global economies and cultures. The internet is based upon open source software (TCP/IP and HTML), which has contributed to its rapid growth and dynamic development. Developing countries, most conspicuously China, now widely use open source software. In the case of China this is manifested in the widespread adoption of the Linux operating system. This signals a slow revolution in the economies of software development and the related intellectual property rights. However, in most developing economies, including South Africa and other African countries, proprietary forms of software predominate. The intellectual property rights are almost entirely held by foreign enterprises, which in effect positions developing country firms as franchisees.

In the case of emerging economies, issues of development are inextricably linked with the manner in which ICTs, including OSS, are managed. For example, by expanding the scope for local innovation, an open source development environment allows local enterprises both to germinate and to move up the international ICT knowledge value chain. The use of OSS is believed to have many further benefits, with the most obvious being lower overall costs. Thus, while OSS is far from a panacea for the developing world's ICT challenges, it arguably has a major role to play.

Policies that aim to leverage OSS for development have been initiated in a number of developing economies, including South Africa. However, in the case of South Africa these policies have been developed in the context of a paucity of relevant data. There are currently no available data describing the extent of OSS uptake in South Africa, whether measured by usage or by expenditure on development. There are no national data with respect to government usage of OSS, the spread of Linux as an operating system, or the human resources employed in OSS systems and development. This means that government policies exist in a measurement vacuum, and since there are no base-line measurements, progress cannot be monitored. Also, a lack of local data makes international comparative research difficult, and it is thus a challenge to situate South Africa in an international context.

This paper therefore offers a quantitative overview of OSS development in South Africa, drawing on recently released primary data from the South African National R&D Survey. This is an annual national survey of all R&D performing organizations in the country. In this instance, software development is included in the 'development' component of R&D, and is therefore measured by this survey. Moreover, since OSS is a national priority area, this form of software development has recently received a dedicated question in the survey. These quantitative findings are supplemented by interview notes from meetings with product development executives from four of South Africa's largest software development firms. Together these findings specify what we currently know with respect to OSS data in South Africa, and indicate what we ought to find out in terms of directions for further research.

This paper is organized into six sections: Section two provides an overview of OSS in emerging economies; Sections three and four discuss OSS policy in South Africa and the issues related to quantifying OSS development respectively; Section five discusses national R&D data for South Africa and analyses R&D effort towards OSS development; and the final section draws some conclusions and makes some recommendations.

2. Open Source Software in Emerging Economies

Grandstrand (2000) describes a world economy that is moving from being industrial to knowledge-based, resulting in the emergence of 'intellectual capitalism'. In this context Information and Communication Technologies (ICTs) play a critical role, as they lower transaction costs associated with information flows and expand the reach of these information flows. ICTs have a particularly important role in the development of emerging economies: information technology supports development, both in the form of increased economic growth and in terms of human welfare (May, 2006). ICT is therefore a necessary component of development strategies for Africa. One of the key debates in the arena of ICTs in development concerns the role of OSS. This debate may be framed as a contest between investing in the current hegemony of proprietary software to drive ICTs, and the increasing recognition that OSS could play a larger role in development. For example, May (2006) reports that in 2006 approximately US$24 billion was transferred from African countries to non-African software companies, mainly based in the USA. As the ICT sector in Africa grows, and as the network predominance of proprietary software continues to grow, it is likely that the scale of such transfers will escalate.

Ghosh *et al.* (2002) identified four key benefits from the adoption of OSS by public institutions. These may be summarized as: (i) lower cost; (ii) independence from proprietary technology; (iii) availability of efficient and low-cost software; and (iv) capacity to develop custom applications and to redistribute the improved products. A move to OSS thus reduces financial outflows to foreign firms, and even in those cases where the short-term costs of OSS development are higher than propri-etary software fees, users of OSS can avoid being locked in to buying future upgrades as well as the management costs of licenses, which reduces costs over the long term (National Advisory Council on Innovation, 2004).

Moreover, the funding required for local development circulates within the local economy, raising its multiplier effect and building human capital in the process. This is one of the key properties of OSS, in that enterprises whose core business is OSS will draw their primary revenue from downstream activities such as distribution and service, rather than income from legally protected intellectual property rights. Thus 'the essence of open source software is not the software. It is the process by which software is created' (Weber, 2004: 56).

In the public debate, proponents of OSS argue that it also supports entrepreneurship, business development, innovation, collaboration, local content and encourages a market for the development of software and media reflecting indigenous knowledge, culture and languages. Many governments, including those of developing economies, have instituted policies supporting OSS (Dravis, 2002). These countries include China, Japan, South Korea, Thailand, Argentina, Brazil, Peru, Germany, France, and the UK. It is widely recommended that 'developing economies and their donor partners should review policies for procurement of computer software, to ensure that options of low cost and/or open source software products are properly considered and their costs and benefits carefully evaluated' (Barton *et al.*, 2002: 105). South Africa's National Advisory Council concludes that OSS should be the 'default exploitation route for government R&D software' (Peeling and Stachell, 2001, quoted in NACI, 2004: 18).

Harnessing the benefits of OSS for development arguably requires that emerging economies must go beyond the use of Linux and other OSS software. Policy needs to be more comprehensive than simply encourag-ing the use of Linux in the public sector, as has been the case in countries such as Brazil and Thailand (Rossi, 2004). It needs to support the skills development that will allow for the local development of relevant applications. In order to attain such benefits, policies need to be based on a thorough understanding of OSS. This includes an understanding of the basic properties of OSS, as well as the incorporation of relevant data into analysis and policy-making processes (Camara and Fonseca, 2007). It is here that this paper aims to make a contribution towards future efforts in measuring, analysing and informed policy making – firstly by outlining the scope of existing data, secondly by drawing tentative conclusions from these data, and thirdly by identifying directions for further research that would help policymakers develop appropriate interventions.

The need for such policy is underlined by the numerous impediments to the uptake of OSS in developing economies. Contextual challenges often include a low absorptive capacity and high failure rates (Ehik-hamenor, 2003). Lobbying pressure from multinational proprietary software firms has a major impact on public procurement. There also remains the question of the impact of legacy software. Replacing such legacy software requires scarce financial and human resources, and also involves an element of risk. Thus, 'without a direct public policy for addressing these limitations, OSS will grow at the fringes of public and private companies and their core applications could remain based on proprietary software' (Camara and Fonseca, 2007: 123). This paper speaks to the necessity of a direct public policy that is supported by sound quantitative and qualitative data.

3. Open Source Software and Policy in South Africa

With the changing of the South African political guard in 1994, the new government initiated an e-government programme that would enable "the continuous optimization of government service delivery, constituency participation, and governance by transforming internal and external relationships through technology, the Internet and new media" (Shilubane, 2001:40). In doing this the national government sought to use information technology as an enabling factor to improve service delivery by automating public access to government documents and services via multi-purpose web portals, personalized services and more efficient delivery of documentation and application processes. Though a number of key stakeholders were consulted in the establishment of the E-Government programme, including the National Advisory Council on Innovation, the State Information Technology Agency, firms and associations from the private sector, and internal government users, the program remains a national government initiative, administered through provincial government structures in all nine provinces.

In attempting to reduce the cost of these programmes, plans were made to save on dollar-based software licensing costs, where possible using local software products. This would also stimulate the local software and ICT industry. The Government Information Officers Council therefore recommended the adoption of Open Source Software as an alternative to proprietary software.

This recommendation galvanized a policy-making process that started with stakeholder consultation and the 2002 publication by the National Advisory Council on Innovation (NACI) of *Open Software and Open Standards in South Africa: A Critical Issue for Addressing the Digital Divide*. This submission to cabinet made Open Source Software a nominal non-negotiable base for the government ICT environment. This document informed South Africa's current Open Source Software policy that in principle is pushing toward adopting an open source operating architecture.

In 2006 a revised policy document was presented by NACI. This continued policy support for OSS in government, but was more prescriptive, highlighting themes, namely *Implementation, Migration, Development, Open Format Licensing, Promotion of The Wider Use of Open Source Software*. Each of these themes defined a particular phase in an implementation plan regarding open source software (see Table 1).

Table 1: South African OSS policy

Policy Statement Theme	Policy Wording
1. Implementation	The South African Government will implement FOSS unless proprietary software is demonstrated to be <u>significantly superior</u>. Whenever the advantages of FOSS and proprietary software are comparable FOSS will be implemented when choosing software solution for a new project. Whenever FOSS is not implemented, <u>then reasons must be provided</u> in order to justify the implementation of proprietary software.
2. Migration	The South African Government will <u>migrate</u> current proprietary software to FOSS whenever <u>comparable software</u> exists.
3. Development	All new software <u>developed for or by</u> the South African Government will be based on <u>open standards</u>, adherent to FOSS principles, and <u>licensed</u> using a FOSS license where possible.
4. Open Content format / licensing	The South African Government will ensure that all Government content and content developed <u>using Government resources</u> is made Open Content, unless analysis on specific content shows that proprietary licensing or confidentiality is <u>substantially beneficial.</u>
5. Promote the wider use of OSS	The South African Government will encourage the use of Open Content and Open Standards <u>within South Africa.</u>

Source: Compiled by authors

What have been the outcomes of these policies? This is a difficult question to answer without corresponding data: while currently available data provide a tentative measurement of OSS development activity, there exists no direct data describing institutional achievement with respect to policy. Gauging the success of these programmes would require monitoring and evaluation (M&E) research. This would require baseline data with respect to government and private sector usage of OSS platforms and applications, the links between these data and government policy initiatives, and qualitative data regarding the impact of these policies. Further to these baseline measurements would be longitudinal data reflecting the impact of these policies over time.

Despite a lack of monitoring data, a scan of institutional progress is a starting point that may provide some informative examples. In 2003 the Council for Scientific and Industrial Research (CSIR) launched the Open Source Centre within Icomtek, which is the information and communication technology hub at the CSIR. This centre had a number of projects assigned under its umbrella, ranging from software development projects to information and awareness dissemination of the benefits of open source among user communities. In May 2005 the Mareka Institute was established, also under the Council for Scientific and Industrial Research (CSIR), with a mandate to support awareness, adoption and development of Free and Open Source Software in research and development as well as implementation initiatives. The State Information Technology Agency (SITA) had at this stage already assisted some government departments with a limited conversion to OSS, but data that might quantify this conversion is limited or unavailable. It is thus clear that a more comprehensive institutional scan would be of benefit for OSS measurement and policy-making.

While government may be taking some action, it is the non-profit sector that has taken the lead in terms of OSS advocacy in South Africa. FOSSFA (Free Software and Open Source Foundation for Africa) was established in Geneva in 2003 in response to the UN and other world bodies calling for more input into African ICT and Open Source development and infrastructure. As its primary mission FOSSFA is mandated to promote the use of OSS in African development projects and coordinate Africa's open source efforts to best utilize available resources and develop new centres of OSS excellence across the continent. In 2004 more than 250 delegates from around the world were gathered at the first FOSSFA-sponsored conference in Cape Town on OSS development for development in Africa, providing some momentum to the OSS movement in Africa.

The Shuttleworth Foundation has been arguably the most significant advocate of OSS in South Africa. Go-open Source, a two-year advocacy drive, was launched in 2004 to promote the use of open source software and provide the public with a resource for information and adoption of OSS technology. The program established a number of advocacy projects that ran until 2006. Tuxlab, another Shuttleworth Foundation project, was initiated in 2002 as a model for setting up computer facilities in South African schools using open source principles and software. However anecdotal evidence suggests that the government has hampered the roll-out of this project by supporting only Windows-based educational software. Other Shuttleworth programmes include the Linux Professional Institute and Hip2b[2], both of which have had a significant (if difficult to measure) impact on OSS in South Africa. This suggests another area of future research that might be of benefit to OSS policy-makers – an examination of the interaction between non-profit OSS projects and government institutions.

4. Quantifying Open Source Software Development

Measuring open source software development is challenging. The most common means of measuring the spread of open source software is by directly or indirectly measuring the level of market penetration of OSS. However, such an indicator is not sufficient for measuring the development of this software – a high level of market penetration may be drawing on OSS developed elsewhere. Other measures must be used to measure OSS development - for example expenditure levels. But such data are rare, and their collection and interpretation face significant challenges.

One such challenge is that a large proportion of OSS development takes place outside the formal working environment, and thus easily escapes measurement. A survey of OSS developers in Brazil found that approximately 40 per cent of developers had a paid job to develop OSS (Stefanuto and Salles-Filho, 2005, quoted in Camara and Fonseca, 2007). A similar proportion was found in the responses to a European survey by Ghosh *et al.* (2002). In both cases 60 per cent of OSS software developers were performing their work outside of the formal business environment, the so-called 'weekend' and 'basement' developers. This finding points to the fact that much of the work within the open source development area is outside of formal industry and therefore may not be effectively tracked and counted in surveys that seek to measure such industries, particularly within the context of developing countries. Therefore any data drawn from the South African National R&D Survey may represent a lower boundary, and possibly only reflect 40 per cent of actual OSS development activity.

A third difficulty is that OSS development falls on the margin of what can be defined as 'software development'. The OECD Frascati methodology (OECD, 2002) describes software development as activity which produces novel software or adaptations leading to novel functionality in existing software. Interviews with executives from four of South Africa's largest software development firms highlighted that this resulted in difficulty in measurement, as much of their development activity would fall on the border of this definition. For example, customization of existing software may introduce new functionality, but classifying this as 'development' requires a careful inspection of the novelty of this functionality and whether this falls within the Frascati manual's definition thereof.

Despite these challenges, there are some available data to describe OSS in South Africa. These data are captured in South Africa's official national annual Survey of Research and Experimental Development Inputs performed by the Human Science Research Council's Centre for Science, Technology and Innovation Indicators (CeSTII), on behalf of the Department of Science and Technology. The survey is carried out according to the standard international Frascati methodology, and the resultant data form part of official statistics for the National Statistics System. Specialised questionnaires are directed at government, science councils, higher education institutions and non-profit organisations (NPOs). The business sector survey employs a purposive methodology, constantly expanding its database of firms that could potentially be involved in R&D in South Africa. Completion of the R&D survey, including basic economic data and extensive R&D data, is a statutory requirement for all organisations performing research or experimental development, including software development.

The R&D Survey is geared towards measuring national priority areas in science and technology. To this end the 2005/06 and 2006/7 R&D surveys requested that respondents estimate the percentage of R&D expenditure allocated to the development of open source software. The firms that answered positively to this question form the sample for this paper.

R&D surveys are inherently imperfect, as they must rely on the co-operation of respondents. While the survey cannot claim to capture all the open source software development taking place within the country, it does include the major players, and is the most complete available data-set with respect to national open source software development. This includes data on expenditure, geographical location, sector, and collaboration.

These data must be seen as a lower bound, and the sample as representative of only those organizations that reported their activity to the national survey. The Science Council and Government sectors have relatively complete coverage, as all the Science Councils and all government departments are required to respond to the survey. Surveying is particularly difficult in the business sector, where firms do not have a strong incentive to disclose sensitive information. This is particularly applicable to open source software development, where international evidence suggests that a large proportion takes place in firms that may be too small to be included in the survey methodology. Individuals or micro enterprises are likely not to be covered by the purposive Frascati methodology.

With these limitations in mind, one can use these data to tentatively describe parameters of the scale, scope and structure of South African open source software development.

In order to provide a qualitative context for this data, the authors undertook interviews with product development executives and management teams from four of the largest software development firms in South Africa. These interviews were part of the extended fieldwork contributing to the National R&D Survey. While confidentiality agreements preclude firm-level details, there were some general findings, which are valuable in setting a context for the R&D Survey data.

5. National R&D Survey Data

The interview findings thus set a context for the quantitative findings. A common point of discussion in all the interviews was that a national ICT skills shortage is having a major detrimental effect on firms developing OSS. The main challenge is attracting and retaining skilled staff. Even qualified staff often have a low level of skills. The root cause of this was identified as a poor quality of mathematics and science skills among those who are qualified. It was noted that because of an emphasis on certification, a South African qualification often does not guarantee applicable skills. Several examples were provided of work being outsourced to India due to a lack of local skills. An extreme shortage of previously disadvantaged staff also reportedly weakens these firms' Black Economic Empowerment (BEE) scorecards, which makes it more difficult to access government contracts and other forms of support. There were complaints about bottlenecks at the ICT Sector Education and Training Authority (SETA), including indefinite delays resulting in an inability to access funding for skills development. The skills shortage is particularly harmful in light of the highly competitive globalised market. A major cause of the underlying skills shortage was seen to be a lack of co-ordination between the Department of Science and Technology (DST), Department of Trade and Industry (DTI), and Department of Education

(DOE), resulting in a lack of common understanding of the problem or the solution thereto. High levels of staff turnover in government reportedly exacerbates the problem.

There was a common response that the policy environment in South Africa lacks knowledge and fails to appreciate the challenges facing local IT firms that strive to be globally competitive. Also, it was felt that the contribution of ICTs to the economy is undervalued in policy circles. As an economic growth enabler it was felt that ICT could make a greater contribution to the South African economy if given more direct policy support (and the implementation of this policy).

The only positive comments with respect to policy were that some of the respondents were making use of the recently instituted R&D Tax allowance, which supports R&D and thus OSS R&D. This may indicate a way forward. One feature of the R&D tax incentive as a supportive measure is that it places the initiative for participation and benefit in the hands of firms. This means that the instrument is less likely to be rendered ineffective by weak implementation by government.

Firm-level interviews highlighted a frustration with government, and available data suggest that government has a minimal level of direct involvement in OSS development. As indicated in Table 2, in 2006/7, out of a sampling frame of 858 R&D performing organizations, a total of 38 were engaged in the development of open source software. The majority of these were private sector firms (26) and higher education institutions (9). Only one government department reported any activity.

Table 2: Profile of sample frame and sample: 2005/6 and 2006/7

	2005/6		2006/7	
	No of organizations performing R&D in the sampling frame	No of organizations performing OSS R&D in the sample	No of organizations performing R&D in the sampling frame	No of organizations performing OSS R&D in the sample
Business	607	20	677	26
Science Councils	24	1	24	1
Higher Education	43	8	44	9
Not-For-Profit	28	1	30	1
Government	88	1	83	1
TOTAL	790	31	858	38

Source: Centre for Science, Technology and Innovation Indicators (CeSTII), Human Science Research Council, South Africa.

However, this is not to say that government policy has been unsuccessful, or that OSS development is not on an upward trajectory in South Africa. The sample of organisations involved in the development of OSS can be placed in the context of general software development data (see Table 3). The 2006/7 National R&D Survey indicates that 141 organisations reported software development activity, representing 16% of the total respondents to the survey. Of these, 73 were dedicated software development firms, in that 100% of their R&D budgets were dedicated to software development. The total national expenditure for software development R&D was R836 million, or 5.1% of total national R&D expenditure. This can be seen in comparison to a total expenditure of R632 million in 2005/6, or 4.5% of national R&D expenditure, indicating that software R&D as a proportion of all R&D has increased.

National OSS expenditure data are reflected in Tables 4 and 5, and again these data represent a lower bound within the Business sector. Approximately R192 million was spent on open source software development in 2006/7, up from R101 million in 2005/6, or a nominal increase of 89%. Whether this increase is part of a sustainable trend or a short-term fluctuation will be determined by the measurements of future national R&D surveys.

Of the 2006/7 total, approximately R118 million (62%) was spent by private sector firms. Government departments accounted for only R4 000, or less than 1%. Overall, the proportion of national software development that was open source was approximately 16.1% in 2005/6, and 23.1% in 2006/7. The two years of available data therefore indicate an encouraging change towards a higher intensity of open source in the arena of software development research in South Africa. Thus not only is software development a growing component of overall R&D in South Africa, but open source has grown rapidly within this sector, mostly driven by private sector R&D investment.

Whether these findings are a result of market forces or successful government policy can only be determined by further research. However, these existing data will be useful as they represent benchmarks that can be used to measure future success in bolstering national OSS development levels – particularly as an outcome indicating the success of national policy.

Table 3: National R&D, Software Development and OSS development profile

	2005/6	2006/7
Total expenditure on R&D	R 14,149,239,000	R 16,520,584,000
Total expenditure on software development	R 632,666,250	R 836,274,259
Software development as a % of all R&D	4.5%	5.1%
Expenditure on OSS software development	R 101,937,110	R 192,785,822
OSS as a % of all software development	16.1%	23.1%

Source: National R&D Survey, CeSTII; *Note:* 1US$ = R8,00.

Table 4: National Expenditure on Open Source Software Development R & D

	2005/6 OSS R&D		2006/7 OSS R&D	
	Total R&D expenditure	%	Total R&D expenditure	%
Business	R 60,476,010	59%	R 118,857,770	62%
Science Councils	R 6,034,520	6%	R 27,509,571	14%
Higher Education	R 27,723,430	27%	R 41,441,391	21%
Not-For-Profit	R 7,700,600	8%	R 4,972,990	3%
Government	R 2,550	0%	R 4,100	0%
TOTAL	R 101,937,110	100%	R192,785,822	100%

Source: CeSTII

The Business Sector survey instrument in the National R&D Survey includes some specialized questions directed at private firms. These include questions about collaboration, which allows for the collaborative propensities of these firms to be represented in Table 6. Contrary to expectations, only 14 of the 26 OSS R&D performing firms reported to be undertaking collaborative activity. However this may be due to incomplete questionnaires being received from some firms. These data suggest that OSS development in South Africa, as elsewhere, is highly collaborative: those firms that did collaborate had an average of 2.7 different modes of collaboration, which indicated that they are part of collaboration networks. The most common collaborative partners were reported to be local associates or partner firms, higher education institutions and science councils. The most common international collaborators were other firms, indicating that some of South Africa's major OSS developers are part of international innovation networks.

Table 5: 2005 / 2006 National R&D Survey: Collaboration profile of firms involved in Open Source Software Development

	Collaboration Mode	Number of firms (n=14)**
South African collaborators	Higher Education Institutions	5
	Science Councils	6
	Government Research Institutes	1
	Members of own company/ Affiliated Companies	6
	Other Companies (including Specialist consultants)	6
	Not-for-profit organizations	2
International collaborators	Higher Education Institutions	2
	Science Councils	2
	Government Research Institutes	1
	Members of own company/ Affiliated Companies	2
	Other Companies (including Specialist consultants)	4
	Not-for-profit organizations	1

Source: CeSTII

The National R&D Survey also includes data describing the geographical location of R&D activity. In line with national trends in R&D expenditure, the main loci of activity were reported in Gauteng and the Western Cape. Thus OSS development in South Africa is not only collaborative but geographically concentrated in the private sector. In essence the core clusters of OSS development in South Africa are collaborative private sector firms operating in Gauteng and the Western Cape.

Table 6: Geographical location of organisations performing Open Source Software Development R & D

2006-2007	Science Councils	Not-for-profit	Higher education	Gov-ernment	Business	TOTAL
Eastern Cape			1			1
Free state			1			1
Gauteng	1		4		17	22
KwaZulu-Natal			1		1	2
Limpopo						
Mpumalanga						
Northern Cape				1		1
North West						
Western Cape		1	2		8	11
TOTALS	1	1	9	1	26	38

Source: CeSTII

6. Conclusions

This paper has highlighted the limitations of our current knowledge about OSS in South Africa. While there exists limited data with respect to OSS development in South Africa (through the national R&D Survey), data describing market penetration of OSS platforms and applications are not available, nor are data describing the extent to which government OSS policies have been implemented. As a result government policy is difficult to monitor or evaluate, making it difficult to determine whether these policies are on the right course.

This identifies some directions for future research that would benefit policy-makers. Firstly, primary research could be directed at government institutions to assess the extent to which OSS policy has actually been implemented – this could be through a more comprehensive institutional scan or through a more detailed follow-up of R&D Survey data. Firm-level surveys would be useful in measuring the market penetration of open source platforms and applications. Longitudinal data would be required for a sustained and reflexive monitoring and evaluation of government policies. Finally, the gathering of primary and secondary data describing the interaction between non-profit organisations involved in OSS advocacy and government institutions would be of benefit in understanding the dynamics between these key players.

However, some conclusions can be reached using currently available data. National R&D Data allow for some benchmarks to be set with which to monitor future performance. For example, national expenditure on open source software development increased by 89% between 2005/6 and 2006/7, from R101 million to R189 million. Approximately 23.1% of all software development reported in the 2006/7 survey was found to be open source, up from 16.1% in 2005/6. South Africa therefore appears to be progressing in terms of Open Source Software development capacity, although additional longitudinal data would be required to assess whether these changes are short-term fluctuations or part of longer term trends. Future national R&D Surveys will allow this comparison and allow for a measurement of the impact of national policies.

National R&D Survey data are indicative of the structure of OSS development in South Africa. OSS development is concentrated in private sector firms in Gauteng and the Western Cape. Organisations involved in OSS development are largely part of collaborative networks, both domestically and internationally. These factors suggest that more detailed firm-level research in these regions would greatly enhance our knowledge of OSS development, and may reveal means to harness the characteristics of these geographical and sectoral clusters, for example through industry organisations, cluster organisations, or other networking structures.

In the context of these data, firm-level interviews have been informative. Key findings were a dissatisfaction with government support for OSS, the identification of a skills shortage as a major impediment to OSS development, and the success of the national R&D Tax allowance in stimulating R&D. Interview-based research with a larger scale and scope may thus be highly informative, and could utilize existing R&D data to identify participants.

References

Barton J., Alexander D., Correa C., Mashelkar R., Samuels G., and Thomas S. (2002), *Integrating intellectual property rights and development policy*, London: UK Department for International Development, Commission on Intellectual Property Rights.

Camara G and Fonseca F. (2007), 'Information Policies and open source software in developing countries,' in *Journal of the American Society for Information Science and Technology*, 58(1): 121-132.

Dravis P. (2002), *A survey on open source software*, San Francisco, CA: The Dravis Group.

Ehikhamenor F.A. (2003), 'Information technology in Nigerian banks: The limits of expectations,' in *Information Technology for Development*, 10(1): 13–24.

Grandstrand O. (2000), 'The shift towards intellectual capitalism – the role of infocom technologies,' in *Research Policy*, 29(9):1061-1080.

Ghosh R.A., Glott R., Kreiger B., and Robles G. (2002)' *The free/libre and open source software survey and study—FLOSS*, Final report, Maastricht, The Netherlands: International Institute of Infonomics, University of Maastricht.

Krogh E., and Von Hippel E. (2003), 'Editorial: Special Issue on Open Source Software,' *in Research Policy*, 32(7): 1149-1157.

May C. (2006), 'Escaping the TRIPs' Trap: The Political Economy of Free and Open Source Software in Africa,' in *Political Studies*, 54(1): 123-146.

National Advisory Council on Innovation, Open Software Working Group (2004), *Free/Libre & Open Source Software and Open Standards in South Africa*, Stanford, California: Creative Commons.

Organization for Economic Cooperation and Development (2002), Proposed Standard Practice for Surveys on Research and Experimental Development Frascati Manual, Paris: OECD.

Peeling N. and Stachell J. (2001), 'Analysis of the impact of open source software,' Government IT Officers Council (Available at: http:/govtalk.gov.uk/library)

Rossi M.A. (2004), 'Decoding the free/open source (F/OSS) software puzzle: a survey of theoretical and empirical contributions,' in *Quaderni dell'Istituto diEconomia*, 424: 1–40.

Shilubane J. (2001), 'e-Governance: an overview,' in *Service Delivery Review*, South African Department of Public Service and Administration, Launch Edition: 40-44 (Available at: http://www.dpsa.gov.za/documents/service_delivery_re view/Launch%20Edition/sdr_launch_ed_complete.pdf).

Stefanuto G.N. and Salles-Filho S. (2005), Impact of the free software and open source on the software industry in Brazil, Campinas, Brazil: UNICAMP.

Weber S. (2004), *The Success of Open Source*, Cambridge, MA: Harvard University Press.

African Journal of Science, Technology, Innovation and Development
Vol. 1, No. 1, 2009
pp. 210-211

Personal Journey from Practice to Innovation Research Scholarship: Reflections

Christopher Freeman[*]

The most important event which influenced all aspects of my life in my early years was undoubtedly the Second World War. It was this war which led to my direct personal interest in technical change and which probably changed the course of my life. The reason for this deep influence was that men like me, who were 18 or 19 years old at the time the war broke out, were waiting to make a decision on the question of which part of the armed services they would serve in. The first year or two of the war were dominated by the success of Hitler's armoured (Panzer) divisions in over-running first Poland and then Norway, Denmark, France, Belgium, Holland and the Balkans. In these campaigns the British Army was involved but with only rather weak armoured forces.

Consequently in 1940-41 when I was waiting for my call up papers and interviews, I decided that I would apply to join a tank regiment or other armoured formations. My wish was granted because the army was itself keen to expand these units and when I was selected for officer training, I was trained as a tank officer at Sandhurst - the army training college which had itself at that time just been converted from training infantry officers to training tank officers.

The training comprised about two months' work operating, driving and maintaining various types of tanks, armoured cars and trucks. I learned to drive a tank before I learned to drive a truck or car and I still prefer it. The second part of the course was on operating and maintaining radio communications during conflict situations. Each tank at that time

---

[*]   Emeritus Professor and Founding Director of the Science Policy Research Unit (SPRU), University of Sussex, Brighton, UK

was equipped with a small radio set, and communication with other vehicles and units played an important role in strategy and tactics. Finally, tanks were equipped with a heavy gun and the officer commanding the tank usually also controlled the use of this weapon.

Later on during the war I was sent on a course on the use of anti-tank weapons – mines and guns of various types, and tactics in using them.

It was therefore military training which led to my interest in technical change. A subsidiary factor was my older brother's enthusiasm for all kinds of technology. Already in the 1930s he operated an amateur radio station with the call sign of G5FZ and in the next attic room to mine he communicated with radio "hams" in all parts of the world. Although I could not do what he did either in electronic or in mechanical technology, I shared his enthusiasm for the international and social aspects of all technologies. The town in which we lived, Sheffield, was deeply affected by the world depression of the 1920s and 1930s and the mass unemployment which devastated the centres of heavy industry. It was the consequence of these two factors – mass unemployment over a long period and the technologies of the Second World War – which together created the personal circumstances that led to my concern with the economics of technical change. The first pieces I ever wrote about economics were on this topic and they have been at the heart of my research and teaching ever since.

African Journal of Science, Technology, Innovation and Development
Vol. 1, No. 1, 2009
pp. 212-219

Innovation in Africa - Toward a Realistic Vision

Bengt-Åke Lundvall*

Less than forty years after Rosenstein Rodan's (1943) seminal article that started the era of 'classical development economics' Hirschman (1981) gave an address where he recognised that the era had come to a close. Hirschman pointed out that he and his colleagues had overestimated the power of their ideas to overcome underdevelopment and specifically he referred to a lack of respect for the emotions and culture of those who were expected to realise the ideas.

I believe that Hirschman was right, not least concerning the African experience, and this is one reason why I see the establishment of a strong research capacity on innovation and economic development within Africa as fundamental for the formation of strategies that can bring Africa out of underdevelopment. And I see the start up of this journal as one important step in this direction. I am therefore happy to respond to Mammo Muchie's invitation to contribute with some ideas for this first issue.

In 2000 I co-organised the Aalborg workshop on innovation systems in Africa with Mammo Muchie, who initiated the project. It was the first time that I was confronted with the challenge to apply ideas about interactive learning and innovation systems to the reality of the African continent (see Muchie *et al.*, 2003). This workshop inspired the formation of the global network Globelics (www.globelics.org) and in 2005, we organised the third Globelics conference at Tshwane University of Technology, Pretoria, South Africa. Globelics is now preparing for the second time in Africa to organise the 7th conference in Dakar, Senegal, later this year. Some of the inspirations for this paper come from

---

* Professor of Economics, Department for Business Studies, Aalborg University, Aalborg, Denmark

browsing the many papers on innovation in Africa proposed for this international conference.

A Realistic Vision is neither a Dream nor a Plan!

Africa has one beautiful and one ugly face. It is a beautiful continent rich in terms of natural resources and amongst its people generosity and friendship are abundant. But both the statistics and a closer look show a different, harsher side of African reality. Millions of children live in poverty and illiteracy, many with serious health problems. Disorder in the form of local wars and crime contribute to the misery of several regions within the continent. Africa has been described as ' the scar on the human conscience'.

Is there any light to see at the end of the tunnel? The harsh reality seems to get in the way of well-intended strategies and plans. Assume that we design a plan based upon 'classical development economics'. We might begin by stating that the only way to grow rich is to industrialise, increase the savings ratio and make major investments in infrastructure, business and people. We might look for 'industrialising industries' and give priority to investment in those. We might recommend that a period of import substitution and regional integration is either combined with or followed by export-led growth.

But then we would have to realise the absence of the most important prerequisites for the success of such a strategy: an efficient state and a culture with strong emphasis on educational attainment and entrepreneurial spirit. The states are not strong and efficient and the culture does not promote the accumulation, diffusion and use of scientific knowledge. This hypothetical scenario seems to confirm the pessimistic reading for Africa.

One of the problems with this analysis may be that its focus is upon *what is missing*. An alternative approach might start from *what exists*. We know for instance that most of the population in Africa still live from agriculture and husbandry. We also know that, not least in agriculture, there is a tension between science-based technical solutions and local traditional knowledge. There is also a tension between the kind of western individualism that spurs private capital accumulation and the collective responsibility in Africa not only for the nuclear family but also for the extended family and members of one's village.

We believe that a *realistic vision* needs to take these facts as starting points – neither wishing them away nor seeing them exclusively as something that should be transformed as quickly as possible in order to create room for respectively 'industry' and 'modernity' in the form of scientism and individualism. But neither can the vision ignore that there is a need for radical transformation of society.

What should be the direction for such a transformation? In this essay, I will start from the assumption that a systemic view with focus upon innovation, knowledge and learning brings insights that are useful when it comes to specifying a realistic vision to overcome the general pessimism about the future of Africa. Rather than setting up a very complex and insurmountable agenda, I will concentrate on a few mechanisms that might open up new avenues of hope.

From the knowledge economy to the learning society

The starting point is that the assumption that knowledge is the most important resource and learning the most important process (Lundvall, 1992; Lundvall *et al.*, 2006) is also valid for Africa. Second, we take into account that important elements of knowledge are tacit and that for all practical purposes the combination of explicit and tacit knowledge is fundamental (Lundvall *et al.*, 2002).

We also recognise that while some learning processes are science-based and make use of codified knowledge, equally important, is experience-based learning (Jensen et al 2007). The most important learning processes are interactive - they require that those interacting with each other have a common language for the communication. To overcome the uncertainty involved in the context of innovation, they also depend upon building mutual trust (Lundvall, 1985).

The competence to learn is critical to any society including Africa. By the competence to learn, we mean learning know-how through experience and through social interaction (Lundvall and Johnson, 1994). Formal education systems therefore need to give access both to relevant codified knowledge and to the competences necessary for life-long learning. The ideal education system is one that combines traditional teaching methods with methods that engage students in the solution of practical problems and that enhance social skills of collaboration and communication (Lundvall, 2008).

Different knowledge institutions may give different relative weights to traditional learning and problem-based learning. The strength of the innovation system will to a certain degree increase with the diversity of its knowledge institutions. In the case of universities diversification may be reflected in a combination of major classical universities (in the case of Africa, one major national or perhaps even only one major regional university in each of Africa's six sub-continental regions), numerous open universities and technical institutions and colleges with strong interaction between higher education and society and, finally, technological institutes without educational tasks and with exclusive focus upon diffusion of knowledge to specialised users.

Innovation may be seen as a process with knowledge both as input and output. The input is codified and tacit knowledge while the output is constituted both by a concrete innovation and the enhancement of the competences of those involved in the interactive learning that is the basis for the innovation. Scientific information and R&D becomes more and more important for innovation in all sectors of the economy, including agriculture. But the same can be said about tacit knowledge and experience-based learning. As distance is suppressed and change becomes more frequent and rapid, individual and organisational learning becomes crucial for technical innovation (Jensen *et al.*, 2007).

Applying the learning society perspective to African development

Enhancing the knowledge base (both scientific knowledge and good practice know-how) for agriculture seems to be the most natural first step for an African strategy for development (Juma, 2008).

Here the innovation system approach is useful because it helps to understand that it is not narrowly focused only on investing in R&D and formal academic training, but also broadly about capability, learning, innovation and competence building systems for development or transformation. The quality of the relationships between knowledge expertise and institutions on the one hand and the competence of the users (farmers and herdsmen) on the other will determine the impact of investments. Therefore, feed-back from users is fundamental for the advancement and efficient use of knowledge.

Therefore a strategy to revitalise the agro-industrial innovation system would involve three elements:

- Investing in the training of scientific and technical expertise that will be of direct and indirect relevance for agriculture.
- Investing in the training of farmers to enhance their capability to absorb and use new knowledge
- Building bridges and networks between the experts and the users.

The three elements are interdependent. To support bridge-building the training of experts needs to have strong elements of problem based learning and strong interaction with farmers during the education processes. Interaction with farmers will help experts to have a realistic idea about the farmers' culture and understanding of the world and will consequently make it possible to communicate 'alien' ideas from science in a form that is respectful and meaningful to the farmers. Land-Grant universities in the US are examples of institutions that combine academic teaching with extension services especially targeted at farmers (Graham, 1994).

Similarly, the training of farmers should aim at making them more open to science, technology and innovation expertise. Early contributions to innovation and entrepreneurship refer to the fact that absorption of innovations was much quicker and more successful in agriculture in regions where farmers had a higher level of education (Nelson and Phelps, 1966; Schultz, 1975).

Ideally the education should offer the farmers positive examples from the use of new methods and consider how the new procedures can be combined with traditional modes of production and distribution. Below we will argue that the Danish historical example of 'People's high schools' may be taken as one source of inspiration for expanding the education of farmers (Lundvall, 2002).

More direct forms of 'knowledge bridges' may be the establishment of local extension service stations that can draw upon formal knowledge institutions. Ideally these institutions should be shaped in such a way that they, without compromising their function to enhance the competence of farmers, are able to take into account local culture.

Who can do it-the State or a new agency?

We may believe that such an agenda should be realised by 'public policy' and by the state. In most positive examples of economic development the state has played a strategic role (Reinert, 1999). But given that the capacity of the state in most parts of Africa is quite limited, we need to consider if there are other agencies that might play a role. The global donor community might play a role (Juma, 2008), but the experience so far is that it has contributed more to foreign aid addiction than to development. While some NGOs contribute positively to development, none of these agencies seem to be able to play a central role in the transformation of the society.

There is a need to establish a new kind of agency. Here we could be inspired by the successful transformation of Danish agriculture in the second half of the 19th century. While the state played a certain role in transforming agriculture from being mainly crop-oriented to being animal-oriented, the most important driver was 'the farmers' movement' (Lundvall, 2002).

The priest Grundtvig formulated the ideology that gave direction to the organisation of farmers into co-operatives. One of the most important elements behind the successful cooperative movement was the diffusion of 'people's high schools' training farmers in economic, societal and cultural issues. They offered not only training in useful skills but also instilled self-confidence among the farmers and prepared them for taking active part in the democratic processes in the society (Lundvall, 2008).

It is difficult to see a positive future for Africa without a similar kind of popular movement based upon the fundamental ideas of diffusing and sharing knowledge. To inspire such a movement, there is a need for a new kind of moral leadership that makes an attempt to reconcile indigenous knowledge with modern science and technology. Hopefully an emerging community of innovation scholars could foster such a leadership in its own midst.

Agriculture and industry

So far we have focused on agriculture. This might seem in conflict with the historical experience that industrialisation is necessary in order to establish sustained economic growth and to abolish poverty. Again Danish history is interesting. While many other economies developed industries based upon minerals and forestry, Danish industrialisation started from agriculture and Danish production and export specialisation remains a reflection of the link between agriculture and industry.

Agriculture constituted an important market for machinery for farm- ing, slaughtering and dairying and as well as for chemical and biological industries. As productivity and income levels increased, markets for consumption good also expanded. The industries refining primary agricultural products became export oriented. Bacon and butter were for almost hundred years the strongest elements of Danish export specialisa- tion. Cooperative ownership of the factories that refine the products was crucial in linking qualitative and quantitative progress in primary agriculture both to market needs and to scientific progress. The gov- ernment supported agro-industrial development by building a spe- cialised university focusing on the needs of agriculture and by financing standardisation institutes.

This strong interlinking and positive interaction between the different elements of the agro-industrial innovation system did not take place by itself. The organisational strength of the Danish farmers' cooperative movement was critically important for the success. In Africa there is a need for a similar driver in order to build the knowledge base both for primary agriculture and for the industries that refine its products and the industries that can deliver its inputs.

Final words

In the realistic vision presented so far, I have overlooked many problems and issues that need to be addressed in order to make Africa a more attractive place to live and work. One fundamental issue is the tension between organising, managing and administrative agencies at and between different levels: local, national, regional and continent-wide. The heterogeneous character of nation states and the potential for ethnic conflicts within them is a strong argument in favour of developing strategies at the transnational level - i.e. at the regional or continental level.

The optimistic view would be that Pan-Africanism as a mission and the African Union as an organisational framework could become vehicles to realise the vision presented here. To exemplify, the attempt to build a 'Europe of Regions' is a modest attempt to link local level interests to the continental in Europe.

We have overlooked other serious problems that need to be addressed. Health problems, environmental problems, including land erosion, child poverty and lack of access to primary education are specific issues that need to be addressed to make life better for the majority. Here we see a formation of a new co-operative agency as a potential vehicle to cope with these problems. An enlightened co-operative movement might be necessary to build and run schools, health clinics as well as to establish local infrastructure.

Not everybody will agree that the vision presented here is 'realistic'. True experts on Africa may characterise it as a rather 'naïve vision'. My counter argument would be that without a vision there is no hope and Africa deserves hope.

References

Graham D.L. (1994), 'Cooperative extension system,' entry in *Encyclopaedia of Agricultural Science*, 1: 415-430.

Hirschman A. O. (1981), 'The Rise and Decline of Development Economics,' in Hirschman A. O. (ed.), *Essays in Trespassing: Economics to Politics and Beyond*, Cambridge: Cambridge University press.

Jensen M. B., Johnson B., Lorenz E., and Lundvall, B.-Å. (2007), 'Forms of knowledge and modes of innovation,' *Research Policy*, 36(5):680-693.

Juma C. (2008), 'Agricultural innovation and economic growth in Africa: renewing international cooperation,' in *International Journal Technology and Globalisation* , 4(3): 256-275.

Lundvall B.-Å. (1985), *Product Innovation and User-Producer Interaction*, Aalborg, Denmark: Aalborg University Press.

Lundvall B.-Å., (ed.) (1992), National Systems of Innovation: Towards a theory of innovation and interactive learning, London: Pinter.

Lundvall B.-Å. and Johnson B. (1994), 'The Learning Economy,' in *Journal of Industry Studies*, 1(2): 23-42.

Lundvall B-Å. (2002), *Innovation, Growth and Social Cohesion: The Danish model*, Cheltenham, UK: Elgar Publishers.

Lundvall B.-Å. (2008), 'Higher education, innovation and economic develop-ment,' in Lin J.Y. and Plescovic B. (eds.), *Higher Education and Economic Devel-opment*, Washington D.C.: World Bank.

Lundvall B.-Å., Interakummerd P., and Lauridsen J. V. (eds.) (2006), *Asia's Innovation Systems in Transition*, London: Elgar.

Lundvall B.-Å., Johnson B., and Lorenz E. (2002), 'Why all this fuss about codified and tacit Knowledge?,' in *Industrial and Corporate Change*, 11(2): 245-262.

Muchie M., Gammeltoft P., and Lundvall B.-Å. (2003), Putting Africa First: The Making of African Innovation Systems, Aalborg, Denmark: Aalborg Univer-sity Press.

Nelson R. R. and Phelps E. S. (1966), 'Investment in Humans, Technological Diffusion, and Economic Growth,' in *American Economic Review* 56(1/2): 69-75.

Reinert E. (1999), 'The role of the state in economic growth,' in *Journal of Economic Studies*, 26(4/5): 268-326.

Rosenstein-Rodan P. N. (1943), 'Problems of Industrialization of Eastern and South-Eastern Europe,' in *Economic Journal*, 53(210/211): 202-211.

Schultz, T. W. (1975), 'The Value of the Ability to Deal with Disequilibria,' in *Journal of Economic Literature*, 13(3):827-846.

African Journal of Science, Technology, Innovation and Development
Vol. 1, No. 1, 2009
pp. 220-248

Research in African e-Local Governance: Outcome Assessment Research Framework

Timothy Mwololo Waema,* Winnie Mitullah,** and Edith Adera***

Abstract

The potential of ICT to contribute to good governance has been known for a long time and has been the subject of many articles and reports, but very little concrete empirical evidence of the effects of ICT on governance, and how these effects should be evaluated, exist. The situation is even worse when we examine ICT and governance local governments. First, drawing upon available literature on e-Government and e-Governance, this paper proposes a research framework for assessing effects of e-Government projects on good governance. In the process, first, it adopts UNDP's nine characteristics of good governance which represent an adequate operational definition of good governance before focusing on three key applications of e-Government. Second, drawing on outcome monitoring and evaluation, the paper proposes a research framework for assessing the e-Governance outcomes as well as a set of good governance outcome indicators. The paper illustrates the usefulness of the framework by providing the generic outcome indicators developed by LOG-IN Africa and outlining a study that was carried out in Kenya following the same framework. Third, suggesting that the proposed framework is a good starting point for investigating e-Governance outcomes, the paper invites researchers to test the framework and create the required knowledge in this area.

Key words: e-Governance, e-Local Governance, ICT, LOG-IN Africa, Kenya, Africa.

JEL Classification: O33, O55, R53

* School of Computing and Informatics, University of Nairobi, Kenya. Email: waema@uonbi.ac.ke

** Institute for Development Studies, University of Nairobi, Kenya. Email: wvmitullah@swiftkenya.com

*** East African Regional Research Co-ordinator, IDRC, Nairobi.

Introduction

The researchers who have worked on e-Governance have come from a variety of disciplines such as computer science, information systems, political science, and public administration. Using theories or frameworks from these disciplines, each comes with his/ her own perspective of what constitutes governance. While the body of knowledge is increasing, there has been no attempt to integrate this knowledge into a solid set of theories or frameworks that others can drawn upon; specifically, in evaluating outcomes or impacts of e-Governance. There is very little that has been produced in terms of a guiding theory or framework.

Instead, positivist statements about the benefits of e-Government and e-Governance dominate the literature. There are claims, for example, that e-Government can lead to the following outcomes: save costs while improving quality, response times and access to services (ADB, 2003); improve the efficiency and effectiveness of public administration (Pacific-Council, 2002); increase transparency in administration, reduce corruption and increase political participation (Seifert and Bonham, 2003); and make governments more competitive (OECD, 2003). Most of these are e-Governance outcomes.

The situation on the ground is not quite so rosy, however, as most e-Government projects or initiatives have not achieved the espoused outcomes. In 2002, for example, the Gartner Group reported that "more than 60 per cent of all e-Government initiatives either fail or fall short of expected outcomes." [1]

In addition, Saxena (2005) argues that in spite of the world wide diffusion of e-Government initiatives, getting the claimed benefits of e-Governance has not been easy for various technological as well as organizational reasons. Heeks (2003) notes too that e-Government projects often fail either totally or partially in achieving their objectives despite their initial successes. Positivist statements or conclusions on the benefits of e-Government and e-Governance initiatives, therefore, have largely been based on studies that have lacked a serious research framework for assessing the associated benefits or outcomes.

The research framework reported in this paper was developed in an attempt to have a guiding model for assessing good governance outcomes of ICTs at the local government level in nine African countries under the Local Governance and ICTs Research Network for Africa (LOG-IN Africa)[2].

In addition, the paper explores the following: a review of the key definitions used in the paper; a description of results-based management by concentrating on the results chain and results-based outcome monitoring and evaluation; focus on the research framework for evaluating the outcomes of e-Governance; illustration of the use of the developed framework with a case study; and finally conclusions and recommendations.

Definitions

Governance and Good Governance

In most economically developing countries, the concepts of governance and good governance have taken root as part of the conditions imposed by the so called "development partners" in development aid as well as being in most cases part of a wider public sector reform programme that is often externally driven through the World Bank and other donor institutions. Even then, the concepts take different definitions depending on the interests of the "donors."

Definitions of governance vary widely depending on the source (e.g. GDRC, 2004a; 2004b; 2004c). All definitions agree on one thing: governance is not synonymous with government. For purposes of this paper, government focuses on activities and outputs by expending effort while governance focuses on processes and outcomes effects produced over a longer time perspective.

Like governance, "good governance" has been defined differently by various authors and institutions. Many definitions are therefore subjective. Given that our assessment framework is on "good governance", it is crucial to make explicit the appropriate and workable definition of this concept.

The World Bank highlights four elements that constitute good governance[3]:

(a) openness and predictability in policy making
(b) professionalism in bureaucracy
(c) accountability of government and participation of civil society
(d) adherence to the rule of law

In addition, the World Bank also provides a list of indicators to measure good governance. However, we consider the elements of good governance as defined by the World Bank as inadequate to take care of the issues we would be concerned about in e-Governance implementation. Firstly, openness is only restricted to policy making. We acknowledge that policy formulation and implementation are highly interrelated and complex and it is often difficult to separate them. However, in e-Governance implementation, we are also concerned about the transparency of the implementation processes and the openness with which key stakeholders have access to information and knowledge. Secondly, participation is only restricted to civil society. We see multiple stakeholders, including private sector and the public, as participants in e-Governance, whether at the policy or implementation levels. Further, it is doubtful whether professionalism in bureaucracy would necessarily ensure efficient utilization of resources while serving stakeholders, an important good governance consideration. Finally, the World Bank definition of good governance excludes concepts that we deem important, including responsiveness and equity.

We now review good governance as defined by UNDP. UNDP describes good governance as being:

> ...among other things participatory, transparent and accountable. It is also effective and equitable. And it promotes the rule of law fairly. Good governance ensures that the voices of the poorest and the most vulnerable are heard in decision-making over the allocation of development resources, and that political, social and economic priorities are based on broad consensus among the three stakeholders - the state, private sector and civil society.[4]

Here, attributes of the concept of good governance are identifiable: (i) adoption of a participatory approach; (ii) transparency and openness; (iii) accountability through assuming responsibilities for actions; (iv) effectiveness, equity and fairness; and (v) endorsement of the rule of law: openness of decision-making to the input of vulnerable social segments as well as the formulation of the national agenda through a consensus between state, private sector, and civil society, all of which serve as the stakeholders in governance.

Most of the definitions of good governance tend to focus on the quality as well as the characteristics of the processes of "governance" (e.g. Weiss, 2000). The definition by UNDP (2002) in which the characteristics of good governance, together with the interpreted meaning, are outlined in the Table 1 below.

Table 1: UNDP's Nine Underlying Characteristics of Good Governance and Their Meaning

Characteristics of Good Governance	UNDP's Definition	Interpreted Meaning
1. Participation	All men and women should have a voice in decision-making, either directly or through legitimate intermediate institutions that represent their interests. Such broad participation is built on freedom of association and speech, as well as capacities to participate constructively"	The act of involving people, regardless of gender, to voice their interests in the decision-making processes. This involvement can be direct or indirect, e.g. the public participating through institutions that articulate their interests
2. Rule of Law	Legal frameworks should be fair and enforced impartially, particularly the laws on human rights"	The act of justly and objectively putting laws in place
3. Transparency	Transparency is built on the free flow of information. Processes, institutions and information are directly accessible to those concerned with them, and enough information is provided to understand and monitor them"	The full release of information and providing stakeholders with free access to institutions, operations and information
4. Responsiveness	"Institutions and processes try to serve all stakeholders"	Institutions and operations answer the requests of stakeholders
5. Consensus Orientation	Good governance mediates differing interests to reach a broad consensus on what is in the best interest of the group and, where possible, on policies and procedures"	Arbitration of the clash of interests in order to establish agreements based on the optimal interests of stakeholders
6. Equity	All men and women have opportunities to improve or maintain their well-being"	The act of providing people, irrespective of gender or other factors of possible discrimination, with equal chances to foster the quality of their welfare
7. Effectiveness and Efficiency	Processes and institutions produce results that meet needs while making the best use of resources"	Achievement of optimal use of resources while serving stakeholders
8. Accountability	Decision-makers in government, the private sector and civil society organisations are accountable to the public, as well as to institutional stakeholders. This accountability differs depending on the organisation and whether the decision is internal or external to an organization"	Managers and decision-makers are held liable to the community
9. Strategic Vision	Leaders and the public have a broad and long-term perspective on good governance and human development, along with a sense of what is needed for such development. There is also an understanding of the historical, cultural and social complexities in which that perspective is grounded"	Managers and the public have a long-term view with regard to governance and are aware of contextual obstacles related to history, culture and society

The list of indicators of good governance often tends to change according to the organisation defining them. We acknowledge that defining the principles of good governance can often be difficult and controversial, but in this paper we think that UNDP's definition of good governance contained in Table 1 would be a good starting point for two main reasons. One, the definition contains a comprehensive coverage of all key characteristics of good governance. Two, the characteristics of good governance have been defined at a fairly generic level, which makes it relatively easier to customize them to the governance context one is handling. For example, there is no mention of any particular stakeholder group in participation. We have thus adopted this definition to include all the relevant factors for evaluation of e-local governance in this research.

The UNDP's definition of "good governance" provides a helpful heuristic, even though it is not necessary to have all the elements of the UNDP's constructs of good governance for an e-Government application to be recognized. The UNDP's definition is wide enough and allows us to investigate different manifestations of good governance in ICT implementations. Indeed, the research teams for whom we developed this assessment framework focused on different elements of the nine constructs depending on the e-Government application they were assessing.

E-Government and E-Governance

E-Government means different things to different people. Ndou (2004:6) identifies three key applications of e-Government:

- "e-Administration" in which an administration or a government office uses ICT in order to interrelate its various departments and digitize its internal operations via "automation and computerization of administrative tasks".
- "e-Citizens and e-Services" which are e-Government applications that enable online access to government information and knowledge and "deliver automated services".
- "e-Society" which provides a platform that facilitates interactions between government actors and civil society.

In this paper, we take Ndou's (2004) broad categorization of e-Government as a good starting point. We shall refine this definition in describing our research framework, however. Like definitions of e-Government, definitions of e-Governance vary greatly depending on who defines it and what are the contexts of its definition. For purposes of this study, we take e-Government to be concerned with the application of ICTs to automate specific aspects of government functions and administrations. E-Governance as a concept is taken to mean beyond e-Government and concerns itself with the impacts of e-Government on governance improvement, effectiveness, efficiency and overall or performance. Thus it not only concerns itself with the mere application of ICTs but also with how these ICTs are used to achieve governance outcomes.

Outcome Monitoring and Evaluation

According to UNDP (2002), outcome monitoring refers to:

> …a continual and systematic process of collecting and analyzing data to measure the performance of UNDP interventions towards achievement of outcomes at country level. While the process of outcome monitoring is continual in the sense that it is not a time-bound activity, outcome monitoring must be periodic, so that change can be perceived. In other words, country offices will accumulate information on an ongoing basis regarding progress towards an outcome, and then will periodically compare the current situation against the baseline for outcome indicators and assess and analyze the situation (UNDP, 2002:20).

How should outcome monitoring be carried out? According to UNDP, to monitor outcomes, managers need to track the outputs and measure their contributions to outcomes by assessing the change from baseline conditions. Further, to conduct effective outcome monitoring, managers need "to establish baseline data, select outcome indicators of performance, and design mechanisms that include planned actions such as field visits, stakeholder meetings and systematic analysis or reports" (UNDP, 2002:20). Monitoring outputs, therefore, is an integral part of outcome monitoring.

Outcome evaluation refers to:

> …an evaluation that covers a set of related projects, programmes and strategies intended to bring about a certain outcome. Such evaluations assess how and why outcomes are or are not being achieved in a given country context, and the role UNDP has played. They may also help to clarify underlying factors affecting the situation, highlight unintended consequences (positive and negative), recommend actions to improve performance in future programming, and generate lessons learned. These periodic and in-depth assessments use "before and after" monitor-

ing data (UNDP, 2002:21).

Monitoring and evaluations of results are performed through performance indicators. Each of the result levels has associated indicators. These are qualitative or quantitative "pointers" that can be measured or described and that help determine progress towards achieving planned results. It is through the regular measurement or observation of these key outcome indicators that we can determine if outcomes are being achieved.

Kusek and Rist (2004) developed a ten-step approach to designing, building and sustaining a results-based monitoring and evaluation system:

(i)   Conducting a readiness assessment

(ii)   Agreeing on outcomes to monitor and evaluate

(iii)   Selecting key indicators to monitor outcomes

(iv)   Collecting baseline data on the indicators

(v)   Selecting the targets for the outcome indicators

(vi)   Monitoring

(vii)   Evaluating

(viii) Reporting

(ix)   Using the findings

(x)   Sustaining the M&E system

In the proposed research framework, we shall take a subset of these steps for monitoring and evaluating good governance outcomes of e-Government projects. The purpose of monitoring and evaluating the outcomes in the case of LOG-IN Africa is to test the usefulness of the proposed research framework, learn some lessons of the effects of e-Government initiatives or projects on good governance and create knowledge that can be used to build a roadmap on e-local governance in Africa.

e-Governance Outcome Evaluation Research Framework

Kettani *et al.* (2005) created a methodological framework for assessing the impact of e-Government systems and governance. The framework does this by assessing the impact of e-Government on good governance outcomes, among other major outcome categories.[5] The researchers used UNDP definitions of the attributes of good governance but customized the definitions to fit the e-Government project that they were studying: a system for providing citizen-related services. They then developed measurable set of indicators for the identified project outcomes. The idea was to assess how project activities and outputs generate project-defined good governance outcomes using defined indicators and compare these with values of indicators before the e-Government system was introduced (or what is commonly referred to as reference or baseline status).

The work of Kettani *et al.* (2005) was our starting point for developing the e-Governance outcome evaluation framework reported here[6] because, like Kettani *et al.*, we took UNDP's definition of good governance as it is inclusive and contains comprehensive coverage of all key good governance characteristics. We also agreed with the researchers on the need to create "working" or contextualized definitions of good governance. Our proposed research framework has two notable differences from the work of Kettani *et al.*

The first point of departure from the work of the researchers under reference is that we have taken into account the various dimensions of e-Government. In our framework, we take the three applications Ndou (2004) defines (as outlined earlier) as reflective of all key dimensions of e-Government. For purposes of our framework, we re-define these components thus:

E-Administration: the computerization of internal government business processes, including business process re-engineering and the development of the required ICT infrastructure;

E-Services: the electronic delivery of government information, knowledge and services to citizens, businesses and other government units through wide area network infrastructure and Internet connectivity as may be required; and

E-Society: the electronic interactions between government actors and citizens, businesses and other government actors. This interaction can be through a number of communication facilities, including mobile phones, cyber cafes and public access facilities.

The key reason we have added this component to our framework is our conviction that good governance outcome indicators will vary depending on the phase an e-Government project is in. For example, the good governance indicators for participation in the e-administration phase of an e-Government project are different from the indicators in the e-services phase of the project. This will become more obvious when we later provide indicators for the various components or phases of e-Government.

The second point of departure from that work is the use of results-based outcome monitoring and evaluation approach to create, monitor and evaluate the good governance outcomes. More specifically, we have borrowed from Kusek and Rist's (2004) monitoring and evaluation steps outlined earlier. We have proposed a *five-step iterative process* as a methodological approach to evaluate good governance outcomes. This approach is outlined below:

1. **Assessment of context**: Conduct an assessment of the context of the e-Government project to gain a deeper understanding of the context of the implementation for the e-Government intervention being studied. The context may include the e-Government/e-Governance policy environment, business and ICT environment, political commitment, ICT infrastructure and connectivity, human resource skills and budget allocation. This context becomes key in explaining why or why not certain outcomes were or were not realized.

2. **Working definitions of 'good governance'**: Create contextualized or "working" definitions of each of UNDP's broad constructs of 'good governance' that the research will focus on. These localized definitions would inevitably be linked to the national government's and/or local government's concepts of good governance as well as the specific context of the e-Government project.

3. **Outcome indicators**: Translate the working definitions of good governance, into identifiable and measurable (or at least observable) outcome indicators. This translation is carried out for each of the three dimensions of e-Government. Outcome indicators can be further specified or associated with identifiable and measurable output indicators which are meaningful in the specific context and directly linked to the e-Government project based intervention.

4. **Baseline data on indicators**: Collect baseline data on the chosen indicators to establish the status of the various outcome indicators before the implementation of the e-Government intervention. This status sets the reference against which future changes can be assessed.

5. **Evaluate the outcomes**: This is the assessment of the status of the chosen indicators to establish the effects of the completed e-

Government intervention on good governance. In general, outcome measurement is a challenge (Mayne, 2007) and indeed this is where the greatest challenge lies. In particular, determining the extent to which an e-Government intervention can be associated with observed outcomes is quite a challenge, largely because of the attribution problem. Further, the long periods from the times outputs are produced to the time "allegedly" associated outcomes are realized creates an additional problem of attribution. In order to deal with these challenges, we propose that researchers focus on the degree to which an e-Government system under study has *contributed* to the status of the outcomes in question, and, at the same time, establish *other factors* that have contributed to the realization of the outcomes.

The above approach is necessary in order to have a sound methodological framework for monitoring and evaluating e-Governance outcomes.

Table 2: Proposed Research Framework

Good Governance Constructs	Working Definitions of Good Governance	Outcome Indicators		
		E-Administration	E-Services	E-Society
Participation				
Transparency				
Responsiveness				
Equity				
Effectiveness & Efficiency				
Accountability				
Rule Of Law				
Consensus Orientation				
Strategic Vision				

Table 2 presents a representation of the proposed conceptual framework, combining the constructs or characteristics of good governance, components of e-Government and the concept outcome indicators across the three aspects of e-Government for outcome monitoring and evaluation. It is to be noted that outcomes will vary from one project to another and therefore a tuning of the constructs of good governance to be more relevant to the project context. This implies the need for a working definition that is more functional, more precise and contextualized to the project under evaluation.

Illustration of the Framework

Generic Outcome Indicators

At a generic level, the LOG-IN researchers collectively developed a set of generic good governance outcome indicators. These are illustrated in Table 3. Each country project team was then supposed to translate these generic indicators into the relevant project-level outcome indicators for that country and in line with the methodological framework outlined above.

Table 3: Generic Good Governance Outcome Indicators

Good Governance Constructs	Working Good Governance Definitions	Good Governance Outcome Indicators		
		E-Administration	E-Services	E-Society
Participation	Stakeholders participate in local government decision-making processes and activities	• Extent of participation in decision-making of staff at different levels	• Extent of involvement of customers, politicians, private sector and civil society in LG e-Governance processes • Extent of availability of electronic delivery facilities empowering citizens to serve themselves without being dependent on employee-intermediaries and their "good will" to do their job	• Extent of improvement in interactions with central government, businesses, citizens and other key stakeholders • Degree of joint planning between local government and civil society • Extent of less privileged people, including illiterate and indigenous people, participating in LG decision making

Table 3: Generic Good Governance Outcome Indicators (continued from previous page)

Good Governance Constructs	Working Good Governance Definitions	Good Governance Outcome Indicators		
		E-Administration	E-Services	E-Society
Transparency	The full disclosure of information, presentation of information in an easy and accessible format, and free access to LG operations and information by stakeholders	• Degree of access and understanding of e-Governance policy and strategy by stakeholders • Extent to which LG business processes and information are clear and accessible to stakeholders • Degree of visibility and trackability of LG internal operations • Extent to which government institutions exchange and share information	• Extent of improvement of public access to pertinent information and services • Degree of ease for citizens to find/ access procedures to follow in order to request and receive services • Extent to which ordinary citizens can read and understand the language by which information and service are presented • Degree of ease for citizens to post comments, questions, and complaints regarding service delivery • Response rate to citizens' comments, questions, and complaints • Degree of availability of government official documents as electronic resources • Extent of availability of the public servant's profile	Not applicable

Table 3: *Generic Good Governance Outcome Indicators (continued from previous page)*

Good Governance Constructs	Working Good Governance Definitions	Good Governance Outcome Indicators		
		E-Administration	E-Services	E-Society
Equity	Services provision or access on equal basis irrespective of gender, disability, socio-economic status or other forms of possible discrimination	• Degree of equity in computerized service access by staff without discrimination • Degree of equity in staff ICT training without discrimination	• Degree of equity in service delivery irrespective gender, country's geographical localization and other forms of marginalization • % reduction in the frequency of "favouritism" incidents in serving customers • Extent of availability of "self-service-technology" (e.g. touch screen kiosks) to empower citizens by enabling them to serve themselves without being dependent on employee-intermediaries as a way to further fight favouritism incidents and endorse the principle of "first come first served"	Not applicable

Table 3: *Generic Good Governance Outcome Indicators (continued from previous page)*

Good Governance Constructs	Working Good Governance Definitions	Good Governance Outcome Indicators		
		E-Administration	E-Services	E-Society
Accountability	LG managers and employees are accountable to citizens and other key stakeholders	• Extent of accountability of LG managers and employees for their actions • Extent of availability of ways to monitor and trace LG personnel activities (i.e. know who does what and when)	• Extent to which LG staff can be identified with their actions in service delivery and be held responsible for them • Degree to which public servants and local governors account to the citizens	Not applicable

Table 3: *Generic Good Governance Outcome Indicators (continued from previous page)*

Good Governance Constructs	Working Good Governance Definitions	Good Governance Outcome Indicators		
		E-Administration	E-Services	E-Society
Consensus Orientation	Conflicts of interest of the various stakeholders are resolved	• % reduction in internal conflicts of interest • Extent of reduction in the possibility of alienation of staff that could be caused by minority versus majority of votes	• % reduction in conflicts of interest in service delivery	Not applicable
Strategic Vision	Managers and the public have a long-term view of governance but are aware of contextual obstacles and their solutions	• Degree of enhancement in long-term view of governance • Extent of existence of e-Governance strategy • Extent of alignment of e-Governance strategy to national socio-economic development priorities	• Degree of enhancement in long-term view of governance • Extent of existence of e-Governance strategy • Extent of alignment of e-Governance strategy to national socio-economic development priorities	Not applicable

As a further illustration of the use of the above framework, we have provided a brief case study of Kenya below. This case study focuses on an integrated financial management system implemented in two local governments.

Case Study of Kenya

National ICT Context

The Government of Kenya, recognized the importance of ICT in 2004, when it launched an ambitious three year (2004-2007) e-Government strategy. This strategy was designed to achieve a set of goals and objectives, namely: to efficiently deliver government information and services to the citizens; to promote productivity among public servants; to encourage participation of citizens in government; and to empower all Kenyans in line with the development priorities outlined in the 2003 – 2007 Economic Recovery Strategy for Wealth and Employment Creation.

The e-Government priorities include: instituting structural and operational reforms; review of the regulatory and legal framework; and development of a reliable and secure infrastructure. The priority activities and implementation framework, over the immediate, medium and long terms include: Government to Government (G2G); Government to Business (G2B); and Government to Citizen (G2C). In G2G, the strategy aims at expanding the information infrastructure, initiating the integration of internal government processes, and to increasing the efficiency and effectiveness of internal operations in the short term. In the medium term (by 2007), the plans include completion of the information infrastructure within the government up to the district level, developing and implementing web-enabled databases, and operationalizing other information systems under development, e.g. civil registration, road transport, revenue collection, etc.

The E-Government strategy further aims at establishing electronic forums for discussion and feedback in G2B and increased involvement of citizens in decision-making and public activities in G2C in the medium term (by 2007). Unfortunately, the E-Government strategy did not mention local authorities. According to the E-Government Permanent Secretary, local authorities were to be addressed through their parent ministry, the Ministry of Local Government. The strategy came to an end in 2007, and its achievements have been evaluated and a draft strategy for the next five years created. By the time of writing this paper, the next e-Government strategy had not been officially sanctioned.

The Kenyan government worked on an ICT policy for several years (Waema, 2005). Indeed, the past years have witnessed significant policy activity in this arena with the finalization of an E-Government strategy, the issuance of a universal access report, the development of two ICT policy drafts, and most importantly, the acceptance of the Government of Kenya of the need for multi-stakeholder participation in the drafting process. The policy was discussed and approved by the Cabinet in January 2006 and an ICT policy document published in March 2006. Interestingly, the e-Government strategy had been created earlier.

The national ICT policy highlights the overall goal of E-Government, which is to make the Government more result oriented, efficient, and citizen centred. The E-Government strategy focused on redefining the relationship between Government and citizens, with the objective of empowering them through increased and better access to government services. The policy gives the broad objectives of E-Government as:

(a) Improve collaboration between Government agencies and enhance efficiency and effectiveness of resource utilization;

(b) Improve Kenya's competitiveness by providing timely information and delivery of Government services;

(c) Reduce transaction costs for the Government, citizens and the private sector through the provision of products and services electronically; and

(d) Provide a forum for citizens' participation in Government activities.

It is hoped that the new e-Government strategy to be soon launched will be in line with the national ICT policy. The current strategy had been developed ahead of the national policy, which was a demonstration of the disjointed approach to policy and strategy. It is also hoped that the new e-Government strategy will highlight the role of LAs in delivering services to citizens and effecting local social and economic development.

Local Governance Context

The Kenyan local governance system is composed of the Ministry of Local Government and the four tiers of LAs, namely: Cities, Municipalities, Towns and County Councils. These councils are corporate entities that are established under the Local Government Act Chapter 265, which is currently under review. In addition to the Act, the LAs draw their legal powers from the Constitution of Kenya, other Acts of Parliament, Ministerial Orders and By-Laws.

While the above legal bodies constitute Local Government in Kenya, the local governance framework in Kenya is broader than the LAs. It consists of provincial, district, location and sub-location administration with technical staff drawn from various ministries. Inherent in the gamut of Local Government and the local governance system are various public and private institutions, including civil society organizations. Kenya has no decentralization policy that rationalizes power sharing, responsibilities, and resources between the central government ministries, parastatals, District Development Committees, LAs and the civil society organisations. This has been a problem at operational local level, with most of the institutions and organizations lacking synergy.

LAs face a number of challenges in realizing their mandate. The challenges include: delivery of infrastructure and services; financial management; institutional and legal framework; human resource capacity; and managing rapid growth. These challenges have resulted in poor service provision and management with many analysts criticizing the LAs, and questioning their role in local development. It is this inefficiency that justified a re-examination of their role and the launching of the LGRP which influenced the LAs to embrace new forms of management including adoption of computerized financial management.

Since the beginning of 1990, when the Government of Kenya began implementing Structural Adjustment Programmes (SAPs), followed by civil service sector reforms in 1993, a number of reforms have been realized. In particular, there has been a deliberate attempt to decentralize governance by moving away from a centralized political system where citizens hardly make any contribution in governance of resources, to a decentralized political system where the inputs of citizens is considered critical to development (Rockeliff and Mitullah, 2003). This has witnessed the review of the Constitution, which stalled, the review of the Local Government Act Chapter 265, which has been on hold pending the completion of the review of the Constitution; review of various Acts of Parliament; the development of Strategic Plans by Government Ministries and parastatals; as well as the adoption of Result Based Management (RBM) by public institutions.

In 1996, a decentralization initiative under the Local Government Reform Programme was launched, it focused on strengthening LAs. The programme has three components: rationalizing central-local financial relations; improving LA financial management, including revenue mobilization; and strengthening citizen participation in planning and ownership of programmes. The reform programme has recognized the importance of LAs in enhancing economic governance, improving public service delivery, and increasing economic efficiency, accountability and transparency (ROK, 1999). The reforms have also included putting in place Fuel Levy Fund, Contribution in Lieu of Rates, user charges rationalization, Single Business Permits (SBP) and the IFMIS. These programs aim at restructuring the local public sector and more import-antly, strengthening local level accountability mechanisms.

KLGRP began with financial reforms aimed at enhancing inter-governmental fiscal transfers, improving financial management, debt resolution, streamlining budgeting systems, service provision, and capacity building for LAs. A key instrument in this process was the enactment of the Local Authority Transfer Fund (LATF) in 1998. The LAs began accessing LATF during the financial year 1998/1999. The Act provides 5 per cent of national income tax to LAs, in line with population, resource base and financial performance. In order to access LATF, LAs are administratively required to develop a Local Authority Service Delivery Action Plan (LASDAP) using a participatory approach. Disbursement of LATF funds depends on strict conditions which if adhered to, would not only improve financial management but greatly enhance accountability and citizens' participation in the affairs of the councils.

Grants from central government include LATF, Road Maintenance Levy Fund (RMLF) and Contribution in Lieu of Rates (CILOR). Prior to setting up of LATF, there was limited sharing of resources with central government. Since the 1999/2000, financial year, LAs have been accessing LATF. The LATF is aimed at complementing the revenue of LAs, which are mainly two: (i) grants from the central government; and (ii) revenue from local sources. Local revenue includes: user charges and taxes from local sources (e.g. property rates); Single Business Permits (SBP); plot rent; market and bus park fees; water and agricultural cess; national reserve and park fees. Municipalities rely more on water and sewage fees, property rates and single business permits while county councils and town councils rely on single business permits, market fees, bus parks and cess revenue.

Implementation of IFMIS

The Integrated Financial Management Information System (IFMIS) is part of the Public Sector reforms focusing on improvement of the public sector, including local governance in Kenya. The Kenya Local Government Reform Programme (KLGRP) is a component of the public sector reform. One of its components is the IFMIS, which aims at improving financial management and revenue mobilization in local authorities. The two cases studies - Mavoko and Nyeri Municipal Councils were the first to benefit from IFMIS in 1999.

Since 1999, the implementation of IFMIS has been extended to six other LAs, namely: Karatina; Embu; Eldoret Municipal Councils; Kiambu; Wareng; and Kirinyaga County Councils. In these councils, IFMIS is used for all financial management activities, including the billing and collection of all local revenues, payroll, and all expenditure controls including issuance of vouchers, procurement, payment of statutory debts and project expenditure management. The financial administrative activities are linked through the budget monitoring system, generating a series of operational and management reports to assist in controlling, monitoring, and managing all financial activities within the LA. The IFMIS began as a national programme and was later broadened to include LAs and hence the concepts Local Authority Integrated Financial Operations Management System (LAIFOMS)[7]. The effect of this system, on good governance, is the subject of this study.

Methodology

Data for this case study was collected from both secondary and primary data. The secondary sources included a review of literature, including government and academic publications, and reports; while the primary information gathering included surveys, key informant interviews and Focus Group Discussions (FGDs). Primary data using survey method was gathered from a sample of staff from the two councils; consumers of the Council services in both councils; citizens and businesses within the jurisdictions of the two councils; and key informants drawn from both councils and central government officials.

In total, 36 council staff questionnaires and 104 consumer questionnaires were administered. Information gathered using these methods was triangulated with outputs from Key Informants interviews and FGDs with senior officers of the councils, councillors and consumers as separate groups.

Good Governance Outcome Indicators

The outcome indicators for the e-Governance assessment study that was carried out for the financial management application in Kenya are shown in Table 4. This study focused on five constructs of good governance[8] in two municipal councils in Kenya (Waema and Mitullah, 2007). The study was carried out by conducting an assessment of the e-Governance context, creating working definitions of the five good governance constructs chosen for this project, creating the outcome indicators pertinent to the project, and evaluating the e-Governance outcomes as per the indicators. Baseline data for the indicators were collected through historical reconstruction as this was a project that had been implemented. The study found that the above framework was useful in assessing good governance outcomes of e-Government projects. More specifically, it found that the integrated financial management system in the two municipal councils was fairly successful in realizing positive good governance outcomes in responsiveness, accountability, transparency and efficiency and effectiveness. Although the study found a significant increase in citizen participation following implementation of the system, focus group discussions established that this change could not be attributed to the system implementation.

Table 4: Good Governance Outcome Indicators in the Study of Financial Management System Implementation in Kenya

Good Governance Constructs	Working Definitions Good Governance (or Outcomes)	E-Administration	E-Services	E-Society
Participation	Stakeholders participate in local government (LG) decision-making processes	• Increased participation of staff in LA decision-making • Increased participation of staff in programme implementation	• Increased gender participation in service access • Increased involvement of stakeholders (customers, private sector, civil society, etc.)	• Improved interactions with stakeholders (central government, business, industry, customers, public, etc.) • Increased inclusiveness in society interactions/ consultations (with business, industry and citizens)
Transparency	The full disclosure of information by LG and free access to LG operations and information by stakeholders	• Extent to which financial policies, processes and procedures are clear and open • Enhanced financial transparency	• Effectiveness of means used to facilitate public access to information • Improved access to information and public services	Not applicable
Responsiveness	Two-way communication between LG officials and stakeholders for requests and feedback	• Extent of feedback from staff and response to the same • Extent of satisfaction with computerized financial services • Community satisfaction index – council's responsiveness towards resolving problems and inquiries	• Extent of feedback from external stakeholders and response to the same • Extent of customer satisfaction with service provision • Extent of access to services by customers	Not applicable

Table 4: Good Governance Outcome Indicators in the Study of Financial Management System Implementation in Kenya (continued)

Good Governance Constructs	Working Definitions Good Governance (or Outcomes)	E-Administration	E-Services	E-Society
Effectiveness & Efficiency	Optimal use of resources	• Enhanced financial performance • Improved budget performance • Enhanced financial management • More efficient and effective delivery of public services • Increased productivity of financial staff	• More efficient and effective delivery of public services • Better quality of services • Extent of customer satisfaction • Reduced cost and time in accessing services	Not applicable
Accountability	LG managers and employees are accountable to citizens and other key stakeholders	• Enhanced financial accountability • Regular and adequate feedback to management on performance of internal control systems	• Extent to which direct service providers can be held accountable for their actions	Not applicable

Conclusions and Recommendations

This paper has viewed good governance as the quality of governance. It has adopted UNDP's definitions of the various qualities of governance as a starting point in defining the concept of good governance, in the process identifying and defining three main components of e-Government: (i) the computerization of internal government business processes; (ii) the electronic delivery of government information, knowledge and services to citizens, businesses and other government units through wide area network infrastructure and Internet connectivity as may be required; and (iii) the electronic interactions between government actors and citizens, businesses and other government actors.

Further, the paper has conceptualized e-Governance as the relationship between the various components of e-Government and a well-defined concept of good governance to produce governance outcomes. With these definitions, it has presented results-based monitoring and evaluation approach as a good basis for investigating the effects of e-Government on good governance. At the same time, it has integrated the defined concepts and proposed a research framework for assessing the good governance outcomes of e-Government projects. We have illustrated the use of this framework with the developed generic good governance indicators at the pan-African level and the specific indicators created using the Kenyan case study.

We believe that the proposed research framework will help researchers to define a set of e-local good governance outcome indicators. This will be done through a process of aggregating, integrating and generalizing outcome indicators from the various national projects to refine the good governance outcome indicators such as those illustrated in Table 3.

In addition, we believe that the proposed framework constitutes a relatively sound research approach to act as a good starting point for empirical studies on the assessment of the effects or benefits of e-Governance. Given the limited knowledge in this area, we invite researchers to use and test the proposed outcome evaluation framework and use case studies to build the required knowledge in this area.

Notes

1   Gartner Symposium/ITxpo, held between April 29 and May 2 at the San Diego Convention Center, San Diego, USA. Available at: http://www. gartner.com/5_about/press_releases/2002_04/pr20020430b.jsp

2   LOG-IN Africa is a pan-African network of researchers and research institutions from Egypt, Ethiopia, Kenya, Mauritius, Morocco, Mozambique, Senegal, South Africa and Uganda. The Network has been carrying out projects to assess the state and outcomes of e-local governance (or the application of ICTs to transform the business of local governments and create governance outcomes) and initiatives. The overall research question is: *"What progress has been made and what are the outcomes in the provision of e-local governance in Africa? What are the challenges and threats? And, what are the good practice strategies and solutions that are emerging?"* More details can be found at: http://www.loginafrica.net

3   The World Bank Group (2006), 'What is Governance? Arriving at a common understanding of 'governance', Available at: http://www1.worldbank.org/ mena/governance/issues-keyQuestion.htm.

4   The Urban Governance Initiative (TUGI) 'Good Governance Issues Report Card: Solid Waste,' Available at: http://www.tugi.org/reportcards/solid waste.PDF.

5   The other major outcome categories were technology, organization, citizen and regulation.

6   The research work done by Kettani *et al.* (2005) at Fez City in Morocco was a key basis for the LOG-IN African project. The work by the same researchers in the City of Larache in Morocco, based on the work in Fez, is one of the projects being carried out under LOG-IN Africa.

7   LAIFOMS is a computer-based system that integrates financial and operational activities involving business transactions in a Local Authority. The system has three main components, namely: revenue, budgeting, and financial management and expenditure components. Each of the components is integrated with each other to have a comprehensive system to monitor all operational activities.

8   In this illustration, the "e-society" e-Government dimension is only applicable to the "participation" good governance construct.

References

ADB (2003), 'To serve and to preserve: Improving public administration in a competitive world,' Available at: http://www.adb.org/documents/manuals/ serve_and_preserve/Chapter19.pdf.

Heeks R. (2003), 'Most e-government-for-development projects fail: How can risks be reduced?,' *IDPM i-Government Working Paper no.14.*, University of Manchester, UK, Available at: http://idpm.man.ac.uk/publications/wp/igov/index.shtml.

Kettani D., Moulin B., and Elmahdi A. (2005), 'Towards a formal framework of impact assessment of E-Government systems on governance,' *Conference Proceedings, the Fourth WSEAS International Conference on E-ACTIVITIES* (E-ACTIVITIES '05), Miami, Florida, USA, pp. 12-20.

Kusek J. Z. and Rist J.C. (2004), Ten steps to a Results-Based Monitoring and Evaluation System: A Handbook for Development Practitioners, Washington D.C.: The World Bank.

Mayne J. (2007), 'Challenges and Lessons in Implementing Results-Based Management,' in *Evaluation*) 87-109.

Ndou V. (2004), 'E-Government for Developing Countries: Opportunities and Challenges,' in *The Electronic Journal on Information Systems in Developing Countries*, 18 (1): 1-24, Available at: http://ejisdc.org.

OECD (2003), 'The e-Government imperative: Main Findings,' in *OECD Observer*, March, Available at: http://www1.oecd.org/publications/e-book/4203071E.PDF.

Pacific-Council (2002), *Roadmap for e-Government in the developing world: 10 questions E-Government leaders should ask themselves*, Los Angeles, CA: Pacific Council on International Policy. Available at: http://www.pacificcouncil.org/pdfs/e-gov.paper.f.pdf.

Republic of Kenya (ROK) (1999), Report of the rationalization and staff rightsizing for effective operation of the Ministry of Local Government, Nairobi: Government Printer.

Rockcliffe-King and Mitullah W. V. (2003), Support to Local Government Programme Components: Output to Purpose Review, Nairobi: DFID.

Saxena K. B. C. (2005), 'Towards excellence in e-governance,' in *International Journal of Public Sector Management*, 18(6): 498-513.

Seifert J. W. and Bonham G. M. (2003), 'The transformative potential of e-Government in transitional democracies,' Available at: http://www.mawell.syr.edu/maxpages/faculty/gmbonham/Transformative_Potential_of_E-Government.pdf.

The Global Development Research Center (GDRC) (2004a), 'Urban Governance,' Available at: http://www.gdrc.org/u-gov/index.html.

The Global Development Research Center (2004b), 'Understanding the Concept of Governance,' Available at: http://www.gdrc.org/u-gov/governance-understand.html.

The Global Development Research Center (2004c), 'Governance: A Working Definition,' Available at: http://www.gdrc.org/u-gov/work-def.html.

UNDP (2002), *Handbook on Monitoring and Evaluation of Results. UNDP Evaluation Office*, Available at: http://stone.undp.org/undpweb/eo/evalnet/docstore3/yellowbook/documents/full_draft.pdf (last accessed 26 June, 2007).

Waema T. M. (2005), 'A Brief History of the Development of ICT Policy in Kenya,' in Etta F. and Elder L. (eds.), *At the Crossroads: ICT Policy Making in East Africa*, Nairobi, Kenya: East African Educational Publishers Ltd.

Waema T. M. and Mitullah W. (2007), 'E-Governance and Governance: A Case Study of the Assessment of the Effects of Integrated Financial Management System on Good Governance in Two Municipal Councils in Kenya,' *ACM Proceedings of the 1st International Conference on Theory and Practice of Electronic Governance*, Macao, China, 10-13 December, pp. 263-268.

Weiss T. G. (2000), 'Governance, Good Governance and Global Governance: Conceptual and Actual Challenges,' in *Third World Quarterly*, 21(5): 795-814.

African Journal of Science, Technology, Innovation and Development
Vol. 1, No. 1, 2009
pp. 249-252

Strengthening Science and Technology Policy Dialogue between Africa and Europe: Understanding a busy landscape and leveraging new opportunities

Daan du Toit*

The following research note was prepared as part of an ongoing work for the FP7 CAAST-Net (Network for the Coordination and Advancement of Sub-Saharan Africa-EU Science & Technology Cooperation) project. CAAST-Net is one of the so-called FP7 INCO-Nets, instruments designed to enhance the EU's S&T cooperation with different international partner regions, notably by undertaking actions such as the identification of R&D priorities ideally suited for cooperation between the EU and the region concerned, and supporting S&T policy exchanges. The CAAST-Net consortium consists of 18 major African and European S&T organizations, including several government ministries, and is therefore well-placed to support the further evolution of the Africa-EU S&T policy dialogue, through providing background policy analysis and other support services.

From the last quarter of 2009, CAAST-Net will convene a series of S&T stakeholder forums to encourage a broader (i.e. beyond intergovernmental forums) African-EU S&T policy debate around a series of key issues of research and development. The EU has allocated 63 million Euros to encourage EU-Africa research partnership.

Science and technology (S&T) cooperation between Europe and Africa at present arguably enjoys unprecedented political attention. This situation can be ascribed to the sustained global focus on supporting African development, in forums of the United Nations, G8, and others;

* Minister Counsellor (Science and Technology), Senior S&T Representative to the EU, South African Mission to the European Union (South African Department of Science and Technology), Rue Montoyer 17-19 B-1000 Brussels.

and also the recognition of the critical role that S&T can play in fighting poverty. In recent years, there has been increasing S&T cooperation between Africa and the European Union (EU). The Africa-EU Joint Strategy adopted at the 2007 Africa-EU Summit in Lisbon for example includes a dedicated Science, Information Society and Space Partnership (the so-called 8th Partnership). The implementation of this partnership has, among other outcomes, resulted in new funding opportunities such as the EU Seventh Framework Programme's (FP7) "Coordinated Call for Africa" supporting African-European research cooperation in areas such as food, agriculture, health, land and water resources.

There is therefore now considerable momentum shaping the bi-regional S&T landscape, and a number of new opportunities for African and European researchers to cooperate. It is, however, important for this cooperation to be complemented by an intensified Africa-European S&T policy dialogue. Within this context, S&T policy dialogue can be considered as the engagement between policy and decision-makers on their respective policies and S&T priorities. The objective is to foster a better understanding of the processes in order to achieve a policy consensus on shared S&T challenges, including the identification of new initiatives for mutually-beneficial research cooperation. Mutual S&T policy learning is also an important part of the process. Policy dialogue is also important, not only to improve the overall enabling environment for research collaboration, but also to link S&T partnerships with political, economic and development cooperation. This will ensure the translation and take-up of research results to support economic growth, social development and other policy programmes.

Prerequisites for a policy dialogue to be effective include not only the involvement of participants with appropriate policy mandates, but also a strongly shared commitment to the process and expectation of its outcomes. The policy dialogue should also be aligned with and embedded within the existing inter-governmental frameworks for cooperation. Within the international S&T cooperation context, a policy dialogue could, for example, interrogate research and innovation responses to pressing global challenges such as climate change, food security, communicable diseases or energy security. It could also constitute a valuable platform for multilateral deliberations on S&T policy issues such as ethics and research, researcher mobility, "brain circulation", the establishment and funding of global research infrastructures, and the international management of intellectual property rights. There are plentiful examples of where the EU's S&T cooperation with international partner regions have been enriched through a strong S&T policy dialogue. These include S&T ministerial forums convened with Asia, and the specific joint steering platforms established with the EU's partner countries in the Mediterranean and Western Balkan regions.

The Africa-EU Joint Expert Group tasked with the implementation of the 8th Partnership is also considering how to initiate a policy dialogue to enhance Africa-EU S&T cooperation. The challenge in establishing such a dialogue is to identify or establish a forum, consistent with the principles elaborated above, which would not duplicate, but complement existing bodies and processes. It should for example be borne in mind that a formal intergovernmental framework for the implementation of the Africa-EU Joint Strategy is provided by the governance mechanisms agreed at the Lisbon Summit (including the Summits, Troika ministerial meetings, and the Joint Expert Groups). There is also regular and fruitful contact at the "services" level between the African Union and European Commissions. Furthermore, in Africa, the African Ministerial Council on Science and Technology (AMCOST), and in Europe the "Competitiveness Council", represent the highest S&T policy-making organs, and these bodies from time to time, engage with international partners through different modalities as and when appropriate .

Whilst a future Africa-EU S&T policy dialogue platform would have to be linked to the Joint Expert Group mechanism, there may be a need for establishing a separate forum. Of course there are existing vibrant Africa-EU S&T policy contacts, taking place at multiple levels and involving different stakeholders, which would also have to be taken into account. Thus, the new platform would have to find its niche among a series of existing dialogues.

To facilitate such a new platform, this note outlines eight different spheres of S&T policy contact between Africa and the EU:

- S&T policy exchange conducted in the formal intergovernmental forums forming part of the Joint Africa-EU Strategy governance framework: this exchange, referred to above, is an engagement steadily gaining in momentum with the implementation of the Science, Information Society and Space Partnership;
- Policy dialogues between Africa's Regional Economic Communities and the EU: within the ambit of their structured development cooperation partnerships, these dialogues increasingly, as for example evidenced by the case of the Southern African Development Community, start to have an interface with S&T policy. However, these are not yet intense engagements;

- Deliberations within the joint forums of the Africa, Caribbean and Pacific (ACP) Group of States and the EU: Although there has not been a new formal S&T engagement since the 2002 ACP-EU Forum on Research for Sustainable Development, the new ACP S&T Programme will, however, undoubtedly rekindle this engagement;
- Africa-EU S&T policy contact within multilateral forums, chiefly those of the United Nations, such as UNESCO (United Nations Educational, Scientific and Cultural Organisation) and the United Nations Economic Commission for Africa (UNECA), including global S&T partnerships on specific themes (e.g. International Panel on Climate Change) or in partnership forums in which both parties participate, e.g. the G8;
- Bilaterally, between an African country and the EU (e.g. those African countries such as South Africa, which have an S&T cooperation agreement with the EU);
- Bilaterally between an African country and EU Member States: There is rich and varied array of bilateral S&T collaborations between African and European states;
- African participations in the FP7: These enable a S&T policy exchange between Africa and Europe (at the level of the participating researchers), whilst the project outcomes provide S&T advice for policy- and decision-making; and
- Regular discussions within global scientific forums, for example of the International Council for Science (ICSU), between African and European researchers, which serve to highlight important policy issues for governments' attention – here the important work of the New Partnership for Africa's Development (NEPAD) Science and Technology Office should be highlighted, there are several NEPAD S&T thematic networks engaging in partnerships with Europe.

The above engagements of course need to be further elaborated and analysed but they are indicative of the rich portfolio of existing engagements, which could support, and on which a new Africa-EU S&T policy platform could be established.

African Journal of Science, Technology, Innovation and Development
Vol. 1, No. 1, 2009
pp. 253-267

Book Reviews

Mammo Muchie*
Abdelrasaq Al-Suyuti Na-Allah**
Angathevar Baskaran***

1. Fetson Kalua, Abolade Awotedu, Leonard Kamawanja & John Saka (eds.) **Science, Technology and Innovation for Public Health in Africa**, DS Print Media, Johannesburg, Republic of South Africa, 2009: pages197.
2. Mario Scerri, **The Evolution of the South African System of Innovation Since 1916**, Cambridge Scholars Publishing, 2009: pages 293.
3. Henk Molenaar, Louk Box and Rutger Engelhard (eds.) **Knowledge on the Move**, International Development Publications, Leiden, Netherlands, 2009: pages287.
4. Gina Wisker, **The Postgraduate Research Handbook** (second edition), Palgrave Macmillan, 2008: pages 428.

1. Introduction

We selected these four books for review together to highlight the interrelated challenges faced by Africa as a whole, by identifying sources that can assist in the search for insights to revitalise research, knowledge,

* DST/NRF Research Professor of Science, Technology, Innovation and Development, IERI, Tshwane University of Technology, Pretoria, South Africa; Professor, Aalborg University. Emails: mammo@ihis.aau.dk; MuchieM@tut.ac.za

** SARCHI/NRF Research Fellow, Institute for Economics Research on Innovation, Tshwane University of Technology, Pretoria, South Africa. Email: abdelrasaq@yahoo.com

*** Senior Lecturer, Middlesex University Business School, London, NW4 4BT, UK. Emails: anga_bas@yahoo.co.uk; anga1@mdx.ac.uk

higher education and doctoral and post-doctoral training. Two of the works are edited volumes. The other two are books. Together they explore in different ways and with different perspectives knowledge production, research capacity, national research and innovation systems, methods for training doctoral and post-doctoral researchers in the context of the challenges posed by globalisation particularly to Africa and broadly to the South.

The first edited volume by Fetson Kalua and others deals with science, technology and innovation for promoting public health care prevention emerging as one of the works accomplished by NEPAD's Office of Science and Technology drawing comparative insights, lessons, education and information for Africa from examples that include Brazil, Cuba, India and other countries. The contributions by seventeen authors in seven chapters cover widely the trends that are influencing public health in Africa based on Africa's Science and Technology Consolidated Plan of Action (CPA) to harness science, technology and innovation by mobilising Africa's human capital and institutions for meeting broad development objectives.

The second work by Mario Scerri on the dynamics of the South African System of Innovation Since 1916 addresses how South Africa, the country that has the internationally recognised status as Africa's current economic power house, has developed historically its national system of innovation in three different periods: 1916 to 1948, 1948-1994, and 1994-to the present. South Africa has a bifurcated national innovation system whilst many African countries have not yet established any functioning systems of innovation. The work is thus important to show how a system of innovation has evolved in an African country historically in order to examine the weaknesses and strengths of this system in light of the need to facilitate the emergence of a few strong and powerful nations in Africa that can serve as useful examples to others. The study can also show how others may be able to forge together or individually their own systems of innovation in Africa to make and consolidate functioning systems of innovations.

The third work which is edited by Henk Molenaar and others explores an attempted critical re-evaluation of the link between knowledge, research and training and the international and national level policy making supporting the process of knowledge production, knowledge movement and brain circulation. Knowledge exists in different disciplinary locations: from single disciplines that are supply-led (Mode I) to interdisciplinary fields that are demand-led (Mode II). A third mode appears to have evolved involving the emergence of peculiarly new patterns of knowledge combinations, where researchers from abroad undertake research defined by those present at the local level. What the editors have described as Mode III knowledge appears to be related to the role in knowledge production of stakeholders such as the donors' in research, the involvement of policy makers, development practitioners, and the twinning of different stakeholders from both mode 1 and mode 2 combining both supply-led and demand-led knowledge production. Fifty authors have put together fourteen essays covering a wide variety of issues such as trade, migration, human rights, agriculture, health, climate, research communication and governance - all these have neatly been captured and encapsulated using the crisp conference title 'knowledge on the move' for the book also.

The fourth research handbook by Gina Wisker deals with the practical advice on how best research training can be undertaken and the steps and procedures that are essential to follow for organising effective post-graduate training and research. It outlines step by step how to go about doing research and attain both consequential and quality post-graduate training and throughput, by providing detailed advice on where and how to start from the beginning, during the ensuing processes, up to and including the completion to the end and beyond. The advice starts how to begin doing post graduate research, by selecting research problems, research methodologies, generating conceptual frameworks, writing a research plan and identifying and matching suitable mentors, carrying out the research by employing suitable methods, theories, cases and literature reviews and crafting and finalising the writing, defending and working through critical feedback and finally moving beyond study and formal training into the professional world of work, scientific development and continuing post-doctoral activities and research. The main purpose of research is to produce the quality researchers who can produce research and train other researchers. It is not simply outputs such as papers, patents, inventions, scientific journals etc. that are relevant, but also the production of quality researchers with economies of scale that can train other present and future researchers and help revitalise the education system as a whole. A research handbook containing ten pages of bibliography (pp. 414-424) with sources of how to create both quality researchers and research is thus critically important to facilitate and encourage in order to fulfil the systematic organisation by improving the completion rate and through close training to generate the much needed quality human capital that is necessary to distribute throughout society and economy, particularly in Africa.

2. The Relevance of the Works to Science, Technology, Innovation and Development

Taken together, each of these four contributions, by applying their own specific perspectives, addresses the significant issues that are relevant ranging from the need for utilising science, technology and innovation for organising an African system of innovation for coping with diseases and facilitate effective and efficient public health delivery; to the historical evolution of the dynamics of innovation systems in South Africa and the challenges of integrating economic development with social development. South Africa is the country that has the most advanced economy in Africa with a real potential, barring all kinds of unforeseen accidents, to emerge potentially as the exemplary and powerful nation in Africa.

The edited volume of 'Knowledge On the Move' brings out the challenges of globalisation and the impact this has for brain circulation for low-income countries that often end up training people only to lose them to areas of the world economy that provides better incentive and remuneration to their graduates encouraging outmigration and brain loss for the countries that need quality trained human capital the most. The various contributors have generated both debate and put forward ideas for new networks and for changing the adverse impact of brain drain into brain gain for the low income countries by strengthening their research systems and research capacity. Models, practices and exemplars are needed that can reverse the flow from the South to the North by making the best in the North to make knowledge, research and researchers move to the most difficult research areas of the world. The Globelics network (www.globelics.org) that links North-South researchers, globally recruited for doctoral training and research flows between the North with the South is one such example that can be replicated to include a variety of disciplines or combinations of disciplinary areas.

The Research Handbook provides the practical guide for effective research training that may be potentially useful to fill the yawning gap and void left after the continuous decline of PhD production in Africa. It is necessary to design and develop well suited research handbooks to discover the best possible ways to train MPhil, doctoral and post-doctoral researchers, particularly in Africa. The attempt to try to reverse the sharp decline of university level training in Africa since the 1980s imposition of structural adjustment, and the impact of this negative policy in shrinking higher education, requires new approaches. There is a need to generate the increasing number and quality of doctoral and post-doctoral researchers and mentors and trainers to feed back to the education, health, university, government, society and economy to accelerate the development process and meet social and economic goals.

Though the four books are dealing with differing themes and issues by each author of the chapters in a given book or between the edited books themselves, it appears that together they inform the reader conveying a shared and broadly interrelated set of concerns. It seems that at the core of their shared concern, however different the approaches, the issues, the views and debates that are straddling the works, is the implicit and/or explicit significance in promoting science, technology and innovation for strengthening a country's innovation system, and building research capacity and research systems that are needed to stimulate and sustain development and transformation, particularly in Africa. Science, technology and innovation can be harnessed and applied to prevent diseases and improve public health delivery in Africa. It is through the spread of science, technology and innovation that the new graduates, the scientists, engineers, innovators and entrepreneurs are weaned requiring the need for the establishment of a self-sustaining and robust system for post graduate training and research in Africa. It is through the strengthening of the science, technology and innovation systems in the context of meeting developmental aspirations that any of the adverse impact of globalisation in stimulating the exodus of brains to the North will be stymied. What emerges is how each of the works reviewed in its own way has tried to relate directly or indirectly how science, technology and innovation can be applied for meeting African development goals.

3. A Brief Review of the Contributions from Each of the Works

A brief overview of the major contributions of the four monographs will be provided by focusing on each of the four one by one.

(i) Science, Technology and Innovation for Public Health in Africa, NEPAD, 2009

The overall organising principle for the production of this edited series of articles on science, technology, and innovation for public health in Africa focused on the exploration of the relationship of innovation systems and public health delivery in the context of Africa's environment.

The objective was to create an African system of health innovation for building shared understanding of how to tap science, technology and innovation to address Africa's burden of diseases. The overall approach of the work suggests that the burden of disease in Africa requires African self-organisation and leadership to deal with it. Outsiders cannot be expected to solve the health problems of Africa. African leadership must take the principal responsibility. African Governments have recognised that investment for science, technology and innovation is necessary for bringing about Africa's structural social and economic transformation. They have transformed OAU into AU and they have added New Economic Partnership and Development (NEPAD) to promote the economic programme and established the African ministerial Council on Science and Technology and separate science and technology offices in NEPAD and the AU. The African Union is on the record to invest 1% of GDP from the member states on R&D that remains unfulfilled to date.

In general quality leadership, predictable institution building and human capital are necessary to streamline science, technology and innovation to promote social and economic transformation. These three important prerequisites are needed to translate declaration into imple-mentation, and rhetoric into practice.

Put together the seven chapters concentrate on the healthcare chal-lenges faced by Africa. As it is often the case when it comes to Sub-Saharan Africa, the statistics, surprise, surprise are all filled with grim pessimism. In this monograph also the following unsettling descriptions abound: human development (lowest in the world), poverty (highest); illiteracy highest in many African countries, spread of HIV/AIDS (cataclysmic), life expectancy low at 46.1 years, AIDS deaths in Africa represent 72% of global AIDS deaths. This monograph follows the same pattern of diagnosis and prescriptions underlined by pessimist temper that others have followed in describing the African predicament whether they start to deal with health care, education, transport, water, agricul-ture, governance or any other issues.

What is interesting about the essays here is the attempt to concep-tualise a health innovation system in Africa to meet local demands for health care provisions. Given the 10/90 gap where only 10% of all health research and development is spent on issues affecting 90% of the world's population, an African innovation health system is something that must be painstakingly organised in order to create an indigenised and embedded health care provision in Africa by being able to take advantage of the global pool of knowledge. Thus building a case for health system research and innovation in Africa, though clearly self-evident, needs to be repeated until all the relevant stakeholders realise it's importance. What is needed also is knowledge building on how such systems of health innovation can be built and sustained given the complexities of the African health, social–economic, research and innovation environment. The chapters in this edition address the relationship of agriculture with health, the relationship between indigenous knowledge and health innovation, the resourcing of health R & D and health innovation and the use of ICT to create and spread electronic health services and the adverse impact of intellectual property regimes in public health delivery especially for the rural poor.

Together the work in this edited volume has managed to highlight the various challenges including the institutional, leadership, resource and regulatory barriers to improve health care delivery. Whilst it is useful to state the various challenges, an important contribution that is needed and to be expected in this volume was how an actual system of health innovation can emerge by considering and taking stock of the fact that in Africa traditional medicine is used by nearly 70-80 per cent of the local mostly rural population. Most of the African population live on agriculture and as health is tied to agriculture how is agricultural transformation to be linked with health innovation for public care? The use of modern ICTs is very important, but given the unreliable infrastructure and electricity supply in the African context, what is the likelihood of using e-health both as a complement and as even one of the means to spread preventive care in Africa? Given intellectual property regimes that continue to shrink health as public good and expand the sphere of health as a private good, what is the appropriate intellectual property regime that is likely to stimulate both a fair and equitable delivery of public health services? What is the chance to change intellectual property with intellectual philanthropy? Can public and private partnerships be organised to expand health as public good and shrink it as a private good or combine judiciously health as a public and private good?

All these issues constitute the challenges in building not just the case for overcoming all the constraints but also for finding ways around the obstacles by building robust and sustainable health innovation and research systems in Africa. The work has brought out the issues for public debate in Africa and further and deeper analyses continue to be in demand given the enormity of the health challenges confronting Africa.

(ii) The Evolution of the South African System of Innovation, 2009

Unlike the NEPAD inspired edited volume above, the work on the South African innovation system is a scholarly contribution that has the theoretical distinction of critiquing neo-classical economics and applying the formal theory of evolutionary economics explicitly to the empirical historical trajectory of South Africa starting from the time the first science and technology plan was launched in 1916 up to 2008. The author divided the dynamics of formation of the national system of innovation into three periods: the first phase from 1916 to 1948 constituted the pre-apartheid phase, the second part from 1948 to 1994 is the apartheid phase, and the period from 1994 to the present is the post-apartheid democratic period. There are continuities and discontinuities between the periods. The author has done these to pin point with precision how much the past weighs heavily on the present and how far can the moves beyond the past be made to clear the ground to undertake contemporary challenges of economic and social development both in South Africa and the rest of Africa.

The author sees the first period as preparing the ground for the establishment of a science and technology base for a 'modernising national economy.' There was a type of segregation preceding apartheid that generated internal conflict. British influence on policy making in South Africa was dominant and the characteristics of the evolution of the national system of innovation was affected by both internal conflict and external dependence on the British colonial empire.

Apartheid as a system emerged in 1948 and until its historical chapter was closed in 1994, it fragmented the society, the economy, the people and the science, technology system, the communities, the universities and the schools. The part that was favoured and supported by the state was building the military industrial complex and Afrikaner agriculture and manufacturing. The majority of the people that were under apartheid segregation remained under conditions of poverty as low income and low wage earners. The South African national system evolved in bifurcated, lopsided and segmented patterns favouring the privileged community over the disenfranchised majority. Apartheid segregation and South African level based and framed national system of innovation for modernising the economy were incompatible. The state abdicated from developing a comprehensive national system of innovation and the science, technology and innovation 'initiatives' that were successful were validated by the same apartheid criteria that initiated and originated them in the first place. "They were mostly driven by the perceived national strategic needs of an increasingly embattled regime and were key constituents of the rapidly growing military-industrial complex" (p.13). Some subsume the evolution of the defence industry to the evolution of the 'minerals-energy complex'(ibid).

Against the backdrop of a history of segregation and apartheid, the new democratic Government that was elected in 1994, launched the White Paper on Science, Technology in 1996 and tried to unravel a national innovation strategy for South Africa. There were two macroeconomic White Papers that the new democratic ANC Government issued: the first was the Reconstruction and Development Programme (RDP) in 1994 and the second was the Growth, Equality and Redistribution (GEAR) White Paper in 1996.

The first social–economic development programme was more Keynesian than neo-liberal whilst the later was more neo-liberal than Keynesian (p. 214). The democratic Government was in a policy dilemma of having to follow economic policy underwritten by the limits on public spending and monetary and fiscal discipline due to the fact that it inherited an economy from the early 1990s with large budget deficits, low foreign exchange reserves, double digit inflation, and a high rate of unemployment on the one hand and to pursue social development programmes to improve housing, water, hygiene, sanitation, electricity, medical services, health, education, land reform and other social services on the other. RDP suggested that government prioritises social policy and GEAR implied it has to privilege stabilising the economy and address the precarious state of the economy. In reality the Government tried to combine both policies; that is GEAR for the economy and RDP for the social realm though not at the pace and scale that could have dealt with decisively with the structural distortion of the apartheid system. Getting macroeconomics right took precedence over removing the inequities and distortions of the apartheid era injustices and relics. The latter has been and is on the agenda but it became something to be tackled, should the fiscally prudent macroeconomic policies and measures work to create jobs. The problem is that formal economy jobs decreased rather than increasing as anticipated. GEAR stabilised the economy but with a jobless growth. GEAR dominated economic policy since the White Paper on Science and Technology was issued, making the latter to have not the expected impact from it.

Paradoxically, the White paper on Science and Technology was framed by using Neo-Schumpeterian evolutionary economics approaches and theories. It used national innovation system to frame the science, technology and innovation policy of South Africa. The White paper was meant to correct the distortions, the fragmentations, the lack of coordination, the lack of national innovation capacity and declining national investment in R&D. The problem was the confusion of following or combining Keynesian Social policy (RDP), neo- liberal and market privileging and state role reducing policies (GEAR) and design science and technology policy within an overall national social-economic policy planning perspective in mind. It was not difficult for government to recognise the importance of science and technology policy, but implementation by identifying specific priorities within industrial and trade policies under a national economic planning framework did not take place. Thus "while the 1996 White Paper was formulated within the context of an entirely different goal set, its proposals on institutional reform failed to address the fragmentation of the S&T planning framework' without altering the definition of the perimeters of science and technology planning" (p. 229).

South Africa had a system of innovation that served a minority. It has to evolve a national innovation system that serves not only all South Africans but also the rest of Africa. It has the largest manufacturing economy in Africa, some of the best universities in the world and a democratic government and institutions. The work on the evolution of the national system of innovation in South Africa is important and addresses critical issues not only by looking back for the sake of remembering history, but also to look far ahead so that South Africa realises fully its potential for itself and the rest of Africa as its neighbourhood. This work is very important to both policy makers and researchers in South Africa, the rest of Africa, and the transition countries and others interested in African innovation for development.

The book is written with 'dense' style and it would have benefitted from the use of language that can facilitate communication and breaks barriers to reach much wide audience.

(iii) Knowledge on the Move, 2009

This collection of essays emerged from a conference on 'Knowledge on the Move' held in the Hague, the Netherlands in February 2008 involving 200 researchers, donors, policy makers and development practitioners. The edited volume brings together the papers presented by the contributors reflecting on 'emerging agendas for development-orientated research.' The key issues addressed by the contributors involved research partnerships, embedding research in society and research capacity development.

The picture that is drawn at the outset by the editors highlights a number of sharp challenges: (a) on the one hand the fragmentation and deterioration of the capacity for research and innovation in low-income developing countries and on the other the concentration and strengthening of research and innovation capacity in the North; (b) the brain drain from poor countries is stark that is illustrated by the example that there are "more African scientists and engineers working in the United States than in the whole of Africa"(p.2); and (c) the internationalisation of research pulls researchers in the developing world towards research agendas that may not prioritise local development. "Research cooperation and development cooperation seem to be moving in opposite directions, one focusing on international research agendas and the other on national development agendas"(p. 3).

Some of the contributors argue for research partnerships that breaks out of the North-South dichotomy with notions of research without borders (Soete, 2009: 33-66). Applied research on health, life sciences, social and human sciences, food, water energy, migration, and so on are neither North or South, they demand research networks and collaborations from both North and South as globalisation/ internationalisation has made them a common challenge rather than a parochial concern to each side to create knowledge for each side alone.

The claim that knowledge generation is moving more to the demand and user end rather than the supply and science intensive end suggests that the ability to translate knowledge into innovation, based on flexibility and trial and error is more relevant than concentrating to re-invent the wheel by trying to embark on original science discovery. An important extension of this shift is to embark on broad based innovation that involve the communities and the local people and NGOs which Prahald (2004) described as Bottom of the Pyramid innovation or some describe as the low hanging fruits of innovation.

The effort to draw the core research from the North as part of research for development can benefit from establishing endogenous innovation policy and research capacity. The concept of how to embed research for development must be combined not merely to diagnose problems but also to find solutions.

The contribution by Berit Olsson suggests why developing countries cannot develop research without creating research communities that can also link with research networks independently with the rest of the world. They need to develop in-country training rather than sending their researchers abroad. The development aid that has been driven by short term concerns need to change and the relevance of long-term and sustained support for research, knowledge and higher education capacity development at institutional and systemic levels is upheld. Structural and not only project focused support are needed, if indeed, strong scientific communities are to be established to contribute to development. A strong case has been made for research universities in poor countries as part of strengthening both research capacity and national research systems.

The rest of the contributions concentrate in looking at and highlighting how research partnerships, embeddeness and capacities are emerging in trade and development, migration and development, human rights with emphasis of rights-based rather than charity-based approach to poverty reduction suggesting stakeholders who contribute to poverty would face legal liability. The chapter on sustainable agriculture and food security brings out success in setting a global network of agricultural research that are also connected to national agricultural research systems. The chapter on health systems and health research draws attention to the disparity between health needs concentration in the South and health research in the North. The authors claim that 95% of the health needs are in the South, and 95% of the health research is focused on the priorities set by the North.

The chapter on climate, energy and environmental care recognised the global focus on the issue. They concentrated in bringing out the water content aspect of the research. They suggest the challenge of climate change should provide opportunity for including Southern networks to participate from setting the research agenda to creating the credible science in the end. The chapter on peace, security and governance argues for research capacity in this area. The chapter on research communication for development underscores the intriguing complexity involved in communicating research as part of the research processes. It is not knowledge deficit that research communication meets through information, it is more what level of understanding and the capacity of stakeholders to use the information that is relevant.

The policy world responded to the input of researchers and research managers and the main input was how to translate emerging agendas for development orientated research into actual policies and their implementation. The call for new understanding of research for development focused on making research partnerships not to be dominated by donor agendas and steered by their power and ability to fund research they choose. More local and Southern ownership was recommended. The donors can support but the research agenda must be driven locally and with local stakeholders. There is recognition of embedding research in society by making stakeholders part and parcel of the research process. There was also recognition for epistemological pluralism and not to give in to positivistic measures of knowledge validity. There is a need to establish long term commitment to build research capacity in the South to overcome donors' priorities of seeing research as a luxury. The Dutch minister of development calls for a global research area, with research networks distributed globally and not specifically in a North American research area or European research area by making donors, researchers and practitioners to develop complementary perspectives. The ultimate recommendation is to complement the changing nature of the development agenda and the research for development agenda by building a new understanding and new relationship amongst the disciplines, the research networks and the stakeholders engaged and implicated in expanding research for development.

The global organisation of research, like the economic distribution in the world, is marked by an unequal and unfair international division of labour. As primary commodities are exported as fuel to the industries of the North, skilled human capital migrates to the North and joins the research areas of the North. This trend can be broken only when strong research areas are built in the South to provide incentives for those in the North to come to the South voluntarily. This situation does not exist yet. What is providing new opportunities not to exclude the South in research is the emergence of one-world issues like climate change. This may provide possibilities to reorganise entrenched cultures and routines for research from the North that may be open to the South.

Knowledge on the move, yes; but knowledge for research on development on the move, not certain it is taking place. The work has brought out some of the most salient issues of research for development. It is recommended that engaging with the issues it raises cannot be postponed.

(iv) The Post-graduate Research Handbook (second edition), 2008.

The task of building a strong research force requires organising training systematically. In Africa donors imposed policies eroded higher education and research infrastructure, undermining both training and research since the 1980s. The cost to Africa of this policy remains still high and it is still affecting Africa negatively. There is a need to overcome this donor driven legacy that still manifests in very weak numbers of quality post graduate trained skilled human capital.

There is thus recognition that the PhD production must be organised and managed by using economies of scale. If this scale economy in PhD production is to be developed, training manuals that can facilitate research should be selected. What is useful about this training research handbook is that it begins with how to start research and goes all the way to Life after Research (p. 410). It is a hands on and 'do's and don't do's' practical guide on a whole variety of issues that arise in the research processes.

There is a need to put together research handbook manuals for making sure that the guidance for the trainees is so organised that the PhD completion rate grows exponentially. This is particularly important to redress the injustices that Africa suffered since the 1980s. Unless the threshold of PhD production is met with speed and resolution by organising high quality human capital, the prospect to create effective research communities would be delayed.

This research handbook is thus very useful for the thousands of PhD candidates that will be joining in African universities in the years to come.

4. Concluding Remarks

The reviews of these four monographs for the first issue of AJSTID have been selected in relation to their contribution to meet the following: the research capacity, research and innovation system, and the search for building post-graduate training, and research and innovation for development needs in Africa.

The making of knowledge on the move in Africa requires appreciating the significance of establishing robust systems of research, training, higher education and innovation endogenously whilst working with networks to benefit from global research. Putting the cart before the horse means to advise Africa to go for global collaboration without establishing robust and embedded research and innovation systems at the same time and even before.

The evolution and dynamics of the South African system of innovation can provide insights on how the making of strong systems of innovation can be forged in the rest of Africa whilst strengthening South Africa's own NSI at the same time.

The construction of a strong African research area is a necessary condition for creating a stronger global research area. In the case of Africa there is no better alternative than to build research and innovation embedded in society and economy and the human capital if indeed Africa is to join the world knowledge economy with equality with the rest of the world. Advice to Africa to join a global research area without first establishing the Africa research area is likely to undermine the creation of strong research communities in equally strong and relevant universities. It is therefore critical that entry into research networks globally enhances the building of Africa's research communities, universities and innovation systems.

Reference

Prahalad C. K. (2004), The Fortune at the Bottom of the Pyramid, Eradicating Poverty through Profits, Upper Saddle River, NJ: Wharton School Publishing.

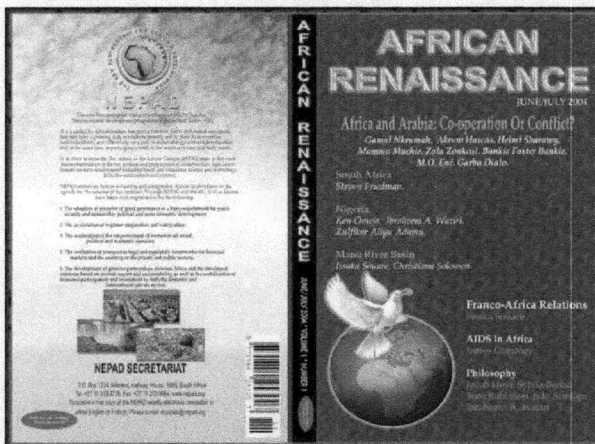

African Journal of Business and Economic Research

AJBER

Annual Subscription Rates

Companies/orgs./institutions: £200
(including access to the online editions)

Individuals: hardcopy only: £50

Individuals: online and Print: £60

Individuals: Online only: £30

Retail sales:
Individuals (print)   £20 (+ P&P)
Online   £10 per issue

For subscription and advertisement enquiries contact:

sales@adonis-abbey.com
Phone: +44 (0) 20 7793 8893

Adonis & Abbey Publishers Ltd
P.O. Box 43418,
London
SE11 4XZ
United Kingdom